Women and Physics
(Second Edition)

Online at: https://doi.org/10.1088/978-0-7503-6435-5

Women and Physics
(Second Edition)

Laura McCullough

Department of Chemistry and Physics, University of Wisconsin-Stout, Menomonie, WI, USA

IOP Publishing, Bristol, UK

ISBN   978-0-7503-6435-5 (ebook)
ISBN   978-0-7503-6433-1 (print)
ISBN   978-0-7503-6436-2 (myPrint)
ISBN   978-0-7503-6434-8 (mobi)

DOI   10.1088/978-0-7503-6435-5

Version: 20240901

IOP ebooks

British Library Cataloguing-in-Publication Data: A catalogue record for this book is available from the British Library.

Published by IOP Publishing, wholly owned by The Institute of Physics, London

IOP Publishing, No.2 The Distillery, Glassfields, Avon Street, Bristol, BS2 0GR, UK

US Office: IOP Publishing, Inc., 190 North Independence Mall West, Suite 601, Philadelphia, PA 19106, USA

*To my husband, Kelly McCullough, who is the most uxorious spouse
a woman could wish for*

Contents

Preface

This book addresses some of the issues that women in physics face right now in the Anglophone world, through a review of the most recent research. It begins with an examination of the numbers of women in physics in English-speaking countries, at all levels of the educational ladder. It moves on to examine factors that affect girls and their decision to continue in science, the issues those girls face as young women in their pre-college physics courses, and the problems they will encounter during undergraduate and graduate studies as they grow older. Moving on from the world of the student, the book outlines the problems that women in physics careers face. The book looks at all of these topics with one eye on the progress the field has made in the past few years and another on those things that we have yet to address. In service of the latter goal, the book surveys the most current research as it tries to identify those strategies and topics that have a significant impact on women's issues in the field with a special emphasis on our biases and stereotypes, and how they can affect the way all of us interact with women in physics both individually and collectively.

Acknowledgements

The author wishes to thank those who assisted in the creation of this work. My husband Kelly McCullough has been amazingly supportive. I appreciate my wonderful groups of friends and family who keep my spirits up. Tracey Berry and Anne Marie Porter helped immensely in multiple ways—thank you! I have many communities that I belong to, and they all have a part in my happiness. (I love seeing my PER colleagues' names showing up in search results.) And I thank the inventor of Inter Library Loan, without whom this book would not have happened.

Author biography

Laura McCullough

Laura McCullough is a Professor of Physics at the University of Wisconsin-Stout. She has a BA in physics from Hamline University, and a MS in physics from the University of Minnesota. Her PhD from the University of Minnesota was in Science Education with a focus on Physics Education Research. She is the recipient of multiple awards, including her university system's highest teaching award, her university's outstanding research award, and her professional society's service award. She is a Fellow of the American Association of Physics Teachers.

Her primary research area is gender and science and surrounding issues, a topic on which she is frequently asked to speak. She has also done significant work on women in leadership, and on students with disabilities. She has had articles published in the *Journal of College Science Teaching*, the *Journal of International Women's Studies*, and *The Physics Teacher*, among others.

Chapter 1

Introduction

A little girl waits patiently at a science exhibit for another child to finish. Her brother joins her and takes over the controls and she never gets her turn.

A young woman in high school physics lab is always relegated to be the record keeper and never gets a chance to play with the equipment.

A woman walks into her first day of physics graduate school and sees twenty-four men and no other women.

A physics professor is called 'Mrs' by her students instead of 'Dr'.

An assistant professor is placed on every departmental committee so they can have female representation.

A woman makes a suggestion at her weekly research group meeting. Her idea is ignored. Three minutes later, a man makes the same suggestion and is applauded.

Women have been doing physics, and loving physics, for hundreds of years. Yet there is still a significant under-representation in the field at every level in the English-speaking world. Why have we not yet reached parity? What issues do women who want to enter physics face? What hurdles do they encounter as they move through university and graduate work? What holds them back in research positions and teaching positions? Why do they leave? These questions have fascinated researchers and scholars for decades.

Why is this such an issue? Should it even matter if there aren't many women in physics?

Yes. For so many reasons.

The presidents of Stanford, MIT, and Princeton know how important it is:

> Until women can feel as much at home in math, science and engineering as men, our nation will be considerably less than the sum of its parts. If we do not draw on the entire talent pool that is capable of making a contribution to science, the enterprise will inevitably be underperforming its potential [1].

Technology and engineering are key areas our nations need to be developing to thrive in this twenty-first century. We should be encouraging every interested child to explore, study, and make a career in physics.

There are many reasons to focus on the complex societal issue of women's participation in physics. The first is probably a simple sense of fairness and justice. Why should some people have an easy time accessing and participating in physics, while others do not? We have ample evidence that there is no biological basis for the lower rates of women in physics; this is a social issue, a cultural issue, a fairness issue. A fascinating example is that of Ben Barres, a transgender scientist who noted how differently he was treated as Ben than he was as Barbara. His autobiography is a wonderful read [2].

A second reason is that our world needs advances in STEM more now than ever. Climate change, pandemics, polluted air and water, food insecurity, poverty: physics plays a part in solving all of these problems. The more people we have working on them, the better and faster our solutions.

Another reason is economic, in the broadest sense. In industry generally, the more diverse the leadership, the higher a company's financial returns compared to average [3]. And it's not just money: Catalyst reports on studies that show improved leadership, lower turnover, and greater employee satisfaction [4]. Interestingly, a higher number of women on executive boards is correlated with a country's math scores for girls on the PISA [5]. Something cultural going on here…

The field of biology tells us that a more diverse ecosystem is a stronger ecosystem. 'Genetic diversity among and within populations of all species is necessary for people and nature to survive and thrive in a changing world [6].' By encouraging more women to enter and stay in physics, we are helping to make our field stronger.

There are a lot of reasons to care about this topic. Given the benefits and importance of promoting women's participation in physics, it is no surprise that national physics organizations have statements about it:

> [The] lack of diversity has both moral consequences; it is a matter of simple fairness and social justice that everyone should have the opportunity to thrive in the physics and STEM community—and economic implications; research shows that better diversity improves outcomes, for both business and the wider economy [7, p 1].

> APS members should work to increase the numbers of members from underrepresented groups in physics in the pipeline and in all professional ranks, with becoming aware of barriers to implementing this change, and with taking an active role in organizational and institutional efforts to bring about such change [8].

> Equity, Diversity, and Inclusivity are foundational values of the [Canadian Association of Physics]. The CAP defines these values as:

1. Excellence in physics is built on the diversity of ideas and people.
2. All members of the CAP will work together to build a culture of inclusion and maintain an environment free from harassment, discrimination, and workplace violence.

> We are committed to fostering an inclusive and equitable environment where all are welcomed and treated with dignity and respect [9].

This volume will explore the participation rates of women in physics, the challenges and hurdles they face, and the things that can encourage and support them in pursing physics. This book is primarily intended as an *introduction* to the topic for both the general audience and academics who may not be familiar with this area of study, though it is the hope of the author that those with some knowledge of these issues will also find material of interest here. Ideally this book will help promote awareness of the barriers women face in entering and remaining in physics, and highlight the importance of pursuing further study on this subject.

It is my intent to present the main issues and topics of women and physics as an overview with enough support to demonstrate their validity. Many of the topics I discuss could be (and have been) the subject of chapters or even whole books. As I was working on this book, I was repeatedly tempted to follow the research into a deep exploration of various subtopics, but I had to remind myself that my goal for this volume was to give readers enough to pique their interest, and start them on their own paths of study.

I hope that this book inspires reflection and action in its readers. We all have a role to play in promoting a more diverse and representative field. Physics needs and deserves the best students and researchers. I want the physicists in generations ahead to be free to spend all their creativity and intelligence on the problems of the day, instead of wasting energy worrying about whether they will be safe, respected, and valued in their chosen field.

'I do not mind that you are a girl, but the main thing is that you yourself do not mind. There is no reason for it [10].' Einstein, to a young girl he corresponded with.

References

[1] Hennessy J, Hockfield S and Tilghman S 2005 Vantage point: look to future of women in science and engineering *Stanford Report* 11 February https://news.stanford.edu/stories/2005/02/vantage-point-look-future-women-science-engineering (Accessed: 27 June 2024)

[2] Barres B 2020 *The Autobiography of a Transgender Scientist* (Cambridge, MA: MIT Press)

[3] McKinsey and Company 2023 Diversity matters even more: the case for holistic impact *Report* https://mckinsey.com/featured-insights/diversity-and-inclusion/diversity-matters-even-more-the-case-for-holistic-impact (Accessed: 7 February 2024)
FCA 2021 Review of research literature that provides evidence of the impact of diversity and inclusion in the workplace *Report* https://fca.org.uk/publication/research/review-research-literature-evidence-impact-diversity-inclusion-workplace.pdf (Accessed: 11 June 2024)
McGregor-Smith R 2017 Race in the workplace: The McGregor-Smith review *Independent*

Report gov.uk https://assets.publishing.service.gov.uk/media/5a7f81c6ed915d74e33f6dc4/race-in-workplace-mcgregor-smith-review.pdf (Accessed: 11 June 2024)

[4] Catalyst 2013 Quick take: why diversity matters *Research* https://catalyst.org/research/why-diversity-matters/ (Accessed: 9 September 2020)
Catalyst 2020 Quick take: Why diversity and inclusion matter *Research* https://catalyst.org/research/why-diversity-and-inclusion-matter/ (Accessed: 11 June 2024)

[5] Noland M, Moran T and Kotschwar B 2016 Is gender diversity profitable? Evidence from a global survey *Working paper* No. 16-3 Peterson Institute for International Economics https://papers.ssrn.com/sol3/papers.cfm?abstract_id=2729348 (Accessed: 30 January 2021)

[6] Hoban S *et al* 2023 Genetic diversity goals and targets have improved, but remain insufficient for clear implementation of the post-2020 global biodiversity framework *Conservation Genetics* **24** 181

[7] Institute of Physics 2022 Written evidence submitted by Institute of Physics *Unpublished by the Institute of Physics (IOP)* FSE0006 https://committees.parliament.uk/writtenevidence/107125/pdf/ (Accessed: 11 June 2024)

[8] APS 2018 Statement on diversity in physics *APS Diversity Statement* https://aps.org/about/governance/statements/diversity (Accessed: 11 June 2024)

[9] Canadian Association of Physicists 2022 *CAP's Equity, Diversity and Inclusion Statement* https://cap.ca/edi/ (Accessed: 11 June 2024)

[10] Gonzalez R 2012 Einstein's advice to women in science still relevant more than 60 years later *Gizmodo* https://gizmodo.com/einsteins-advice-to-women-in-science-still-relevant-mor-5900385 (Accessed: 27 June 2024)

IOP Publishing

Women and Physics (Second Edition)

Laura McCullough

Chapter 2

Important notes about this book

The greatest enemy of truth is not error, but prejudice.
—Ashley Montague, *The Natural Superiority of Women* [1]

Working on the second edition of this book in 2024, the world is a very different place from 2016 when I initially wrote it. A pandemic shut down public spaces all over the world; political reaction to vaccines and other medical interventions created or exacerbated serious divisions in some societies; reactionary and intolerant leaders have been elected to many offices.

Similarly, but more positively, the research literature on women and physics is in a very different place. Instead of gaps, gaps, gaps, we now focus on the culture of physics and facets that support or inhibit women taking their place in physics and other STEM fields. You'll read about developing or creating a sense of belonging and having a scientist identity. We continue to learn about how families and early school experiences can affect the choice of physics in secondary or tertiary education. We unfortunately haven't made a lot of progress in the numbers, but cultural awareness of the issue and the number of programs and interventions supporting diversity in physics have increased dramatically. Allies, bystander intervention, and pronouns are things we didn't much consider in relation to this topic even ten years ago.

Because there is so much that is different, I want to make readers aware of how I approached this work, and how they may wish to interpret what they find in these pages.

2.1 Anglophone focus

As with the first edition, I limited the scope of my work to English-speaking countries. My focus was on the US, UK, Ireland, and ANZAC. One reason was that these societies have much more in common than, for example, the US and Germany or India or China. Second, even though English is the *lingua franca* for most of science, I didn't want to be in the position of attempting to translate research and possibly losing important meaning and context about a study. Third? I'm American. I know my own culture and the research happening here the best. It is the area in which I am most confident of my content area expertise. As an adjunct to an awareness of my own strengths and weaknesses, I must also note that I suspect my databases have an American bias.

If you want to start learning about women in physics in other countries, the conference proceedings from the triennial International Conferences on Women in Physics have country-specific papers as well as more general papers. Available for free at AIP Publishing [2], they are both informative and fun to read!

2.2 Physics, science, STEM

Since this book is about women and physics, I specifically looked for literature that focused on physics. In places where there is none or it is extremely limited, I moved to a broader search for science. And if that failed, I moved to the broadest category of STEM. But, as intended, the research is focused narrowly on physics wherever I had that option. When I did have to cast a wider net, I have tried to be mindful of mentioning when the focus was broader, using 'science' or 'STEM' as appropriate. For primary school, the research is almost all broadly 'science', with physics entering at the secondary level.

One significant reason for focusing tightly on physics is that when we look at 'science' more broadly, it typically includes biology, which is mostly female dominated. The issues for women in biology are very different than for women in physics. And when you open it up to all of STEM, the research may include the social sciences, which again are mostly majority women. Engineering is closer to physics than other sciences (culturally and by the numbers), and in places I have noted research in that area.

With the exception of the chapter on history, I have chosen to highlight articles in the last 10–15 years, unless there is a seminal or landmark paper readers should be aware of.

One significant bias in the research literature is it centers studies on students and academics. This is a *major* bias. It was quite difficult to find research on industry and career issues. I've done as well as I could, while still maintaining the focus on research, and not branching out into the more subjective areas of memoirs and essays.

2.3 Gender, sex, gender identity, gender presentation

Most of my readers will have encountered someone who has a gender identity, gender presentation, or pronoun that does not match their sex at birth. (If you haven't in today's environment, you should definitely be working to expand your networks!) A nice overview of sex, gender, orientation, etc, can be found by searching online for the 'Genderbread Person'. It's good for those who are still figuring out the language, culture, and social implications of this area of human experience, as well as for those who want to be sure they are keeping up with current language choices. For the sake of language and style, I will use female and woman mostly interchangeably, even though we more typically think of female as sex and woman as gender.

When I talk about 'women', I mean individuals that have given their preferred gender identity to the researchers, their names/pronouns/pictures were used to make a reasonable guess about their gender by researchers, or (most commonly) checked a box that said 'woman'. I have not chosen to note how gender was determined in every study, as methodology is varied and not central to the topic of this work. Readers can dig into cited literature as they wish.

Nearly all of the references in this book involve research using a gender binary: male or female, woman or man. The number of nonbinary, trans, genderqueer, etc, people in physics is probably small as a percentage of total physicists, though we don't actually know what the data look like in any detail. For reference: the IOP survey of members in 2019 lists 83% male, 17% female, and 1% other, with 2% having a different gender identity than at birth [3]. An APS member survey in 2015 had a 22% response rate and 2.5% of respondents identified as LGBT+, but another 14% declined to provide this information [4]. How many of those identified as trans is unknown. A 2022 Pew report says 1.6% of US people identify as trans or nonbinary [5].

There is almost no research in physics on nonbinary and trans people; hopefully this will start to change soon! (A poorly phrased web search on 'physics nonbinary data' offered up a few mathematical papers.) One paper I found in astrophysics has a nice summation of arguments against the way in which gender-related research has been done in the past:

> We identify three major concerns common to most or all of the analyses
> … (1) the treatment of gender as observable through means other than
> self-identification; (2) categorization schemes with limited gender
> options; and most critically, (3) an over-reliance on quantitative method-
> ology [6].

A great resource in this area is the paper 'Sex and gender as nonbinary: what does this mean for physics teachers?' [7] And there are reports from IOP and APS on LGBT+ populations [8].

Because both the numbers and percentages are small, large sample research studies often ignore the issue, requiring respondents to choose male or female, or they toss out data from anyone who doesn't fit those two categories. Given sample size issues, the best way we'll be able to understand these issues is by doing case studies and ethnographies for those in this select population.

2.4 Women, or women versus men?

For most of the history of women in physics research, the focus has been on the differences between men and women. There are certainly problems with this approach. Typically, men have been the standard and women are judged against them, producing 'deficit' type research, where women are framed as less than/worse at/lower than men. This is also the structure of a lot of the 'gap' research where researchers looked for the gaps between men and women on performance, attitude, or a host of other things.

Along with issues of intersectionality (below), there is a growing push for studies that don't compare women to the arbitrary standard of men. A nice argument is made by Traxler *et al* on this topic [9]. Studying women's experiences has important value in and of itself. This volume has both kinds of research.

2.5 Intersectionality

One of the biggest booms in the research literature since the first edition of this book is work that addresses the idea of intersectionality. Initially coined by Kimberle Crenshaw in 1989, this is not a new concept [10]. But it has really come into its own in the last decade or so for gender research.

Intersectionality is the understanding that every person has multiple identities and they all must be considered when examining ideas of privilege, discrimination, and marginalization. With an initial focus on Black women (race + gender), intersectionality has opened up huge opportunities for research. A cisgender White woman has a different experience than a cis Black woman, or a biracial neurodivergent woman, or a White lesbian with a disability, or a first-generation Latina. We cannot come to a complete understanding of the factors affecting women in physics until we start exploring how women's multiple identities interact with the culture of physics. A good start to understanding this concept is the book by Collins and Bilge [11].

Because the work in intersectionality for women and physics is still relatively new, the related literature sparse, and the nature of this book is to be a broad introduction, the studies in this book are almost all framed around gender only. Few also dig into issues of race or disability or sexual orientation or socioeconomic status. Where research is available, I include some studies in the appropriate places, and if there is ever a third edition, I would hope the progress in this area will be such that I will have the opportunity to include much more work along these lines.

For a good place to start on race and physics, try the AJP 'Resource letter RP-1: race and physics' [12]. This includes a section on intersectionality. And definitely look at Maria Ong's *The Double Bind in Physics Education* [13].

For information on LGBT+ members of physics, you definitely should read *LGBT Climate in Physics: Building an Inclusive Community* [14]. In a study on STEM individuals conducted in 2021, the authors conclude that this community is strongly impacted by inequity in STEM [15]. This is an under-researched population.

2.6 *Hors d'oeuvres* versus the plated meal

Readers, please keep in mind that this book is intended as an introduction to the many facets of the issues physics faces with regard to women and other minoritized groups. I am not going into depth on any of them, but hope to give you enough information and reference material to pique your interest and send you off to do your own digging into the literature. There are many, many papers of interest and import that did not make it into this volume. No disrespect is intended to those scholars; rest assured that vast numbers of those papers are in my personal files and of great interest to me as a researcher! I must also note that I am one person using a single primary database (EBSCO) and operating in a limited time frame. Searches typically returned many, many medical physics titles which, while they looked quite interesting, fell outside the intended scope of this book. I know that I missed things; I hope that the lacunae are not frequent or large.

2.7 Emotions

If I have done my job properly, you will find things in this book that will surprise you, please you, and anger you. It is important to remember that learning is absolutely an emotional task; we remember better when our emotions are engaged. (I learned a lot from the book *The Spark of Learning: Energizing the College Classroom with the Science of Emotion* [16] and can recommend it to any teacher!)

If you find yourself getting upset, take a break. Go enjoy a nice walk or pet a friendly animal, and come back later.

If you find yourself getting angry, use that energy for good! Offer to go into a classroom to talk about this important topic using the resources from the US AIP [17]. If you can, buy some books on women and science for your local lending library, little free library, or school. Share stories and links on social media. Boost the profiles of the women doing amazing work.

If you find yourself getting depressed, come back to this page right here and read the following.

Things are so much better than they were. My mother-in-law's experience being discouraged from science was far removed from my chilly experience in grad school, and they both are far removed from what today's young women will face as they aspire to be part of this exciting field.

I was recently listening to an episode of the quiz show *Wait Wait Don't Tell Me* with feminist icon Gloria Steinem as the guest [18]. The host, Peter Sagal, mentioned that Ms Steinem had been doing this a long time and asked 'Do you ever find that young people today, especially even young women, have a hard time believing how terrible it was when you started your particular journey?'

Her answer was very similar to the one I give when I am asked the same sorts of questions while giving a talk. She said 'Yes, I do, and I'm so glad!' She continued: 'We who are older are hopeful, because we remember when it worse, and young people who don't remember when it was worse are mad as hell because it's not better now.' And the clear implication was that those younger people would be doing the work to make it better.

Things are so much better than when I started doing this research. The numbers still need to move, but attitudes, support structures, climate … it's all so much better, and that gives me great hope.

We have come to a solid understanding that the issues are entirely cultural, and we know cultures can change. Every person in our broader culture has a role to play in improving the chances for women to succeed in physics. You can be an ally, learn how to intervene when inappropriate things happen, give support to someone struggling, be a visible face of someone who thinks women should be in physics, share on social media, and do many other things to make this field a better place. It doesn't matter if you're a physicist or a woman or a fuzzy alien from Koozebane. Everyone can help.

2.8 My biases

No one enters into a research project as a blank slate. Our identities, our experiences, and our intentions all play a role in how we do research, whether it is examining high energy particles or high energy students. I come to this book with a lot of career experience researching women in physics, but also over 30 years of personal experience as a woman in physics.

I have a physicist father, moms in chemistry and math education, grandparents in education, a sister in education, and a brother with a physics degree. I have wanted to be a physics professor since I was in high school. I loved my small college undergraduate experience, with three women among the six or seven physics majors. I got to graduate school and found myself the only woman in my class. The climate in the department was chilly, and I didn't feel welcomed or encouraged by more than a few faculty.

After I completed my master's, I switched to an education PhD because I knew I wanted to teach, it was a much more friendly department, and I didn't see how advanced physics coursework or research would help me be a better physics faculty member. I love the physics education research (PER) community. It is also the place where a lot of research on gender and physics is published.

Because of my graduate school experiences, I started looking into issues involving women and physics. Over the years, it has broadened to women and science, women and STEM, and women and STEM leadership. It really should be gender and science, and usually I speak about my specialty that way. But for research, unfortunately, it's still a very gender binary field, and it really is about women.

I am a cisgender White woman from an upper-middle-class background, and this gives me a lot of privilege despite being a minority within physics. I try very hard to be conscientious about treating those with less privilege with respect, listening to (and believing) their stories about their lived experiences. I wish I could go into more detail about all the intersectionality research that is out there in physics, but that would be another several chapters, and falls outside the main scope of this work. I bring it up here and there to remind people that we can't really understand issues for women in physics until we thoroughly explore issues *among* women in physics.

My biases will be present in this book, despite my efforts to address and ameliorate them. At one point, I found myself focusing on gender differences instead of differences *and* similarities, and had to explicitly remind myself that I should not be leaving out any research that is well done and meets the intended scope of the book. For much of the performance data, the differences within a gender are much bigger (and more important) than differences across genders [19].

I am extremely grateful to all my friends, colleagues, and to the many, many researchers cited in this volume. Their help and their work has been invaluable, and any mistakes are my own.

The goal of this book is to educate and inspire. I hope when you finish it, you will feel that I have succeeded.

References

[1] Montague A 1999 *The Natural Superiority of Women* (Lanham, MD: Rowman and Littlefield) p 53

[2] AIP Conference Proceedings https://pubs.aip.org/aip/acp

[3] IOP 2020 We are physicists *Results of the IOP Member Diversity Survey 2019* https://iop.org/sites/default/files/2020-11/IOP-Diversity-Inclusion-Report-AW-Nov.pdf (Accessed: 30 June 2024)

[4] Atherton T, Barthelemy R, Deconinck W, Falk M, Garmon S, Long L, Plisch M, Simmons E and Reeves K 2016 *LGBT Climate in Physics: Building an Inclusive Community* (College Park, MD: American Physical Society)

[5] Brown A 2022 About 5% of young adults in the US say their gender is different from their sex assigned at birth *Pew Research Centre Short Reads* https://pewresearch.org/short-reads/2022/06/07/about-5-of-young-adults-in-the-u-s-say-their-gender-is-different-from-their-sex-assigned-at-birth/ (Accessed: 17 June 2024)

[6] Rasmussen K, Maier E, Strauss B, Durbin M, Riesbeck L, Wallach A, Zamloot V and Erena A 2019 The nonbinary fraction: looking towards the future of gender equity in astronomy arXiv: 1907.04893 5

[7] Traxler A and Blue J 2020 Sex and gender as nonbinary: what does this mean for physics teachers? *Phys. Teach.* **58** 395–8

[8] IOP, RAS, RSC 2019 Exploring the workplace for LGBT + physical scientists *Report* https://rsc.org/policy-evidence-campaigns/inclusion-diversity/surveys-reports-campaigns/lgbt-report/ (Accessed: 17 June 2024)
Atherton T, Barthelemy R, Deconinck W, Falk M, Garmon S, Long L, Plisch M, Simmons E and Reeves K 2016 *LGBT Climate in Physics: Building an Inclusive Community* (College Park, MD: American Physical Society)

[9] Traxler A, Cid X, Blue J and Barthelemy R 2016 Enriching gender in physics education research: a binary past and a complex future *Phys. Rev. Phys. Educ. Res.* **12** 020114

[10] Crenshaw K 1989 Demarginalizing the intersection of race and sex: a black feminist critique of antidiscrimination doctrine, feminist theory and antiracist politics *Univ. of Chicago Legal Forum* **1** 139–67 http://chicagounbound.uchicago.edu/uclf/vol1989/iss1/8

[11] Collins P H and Bilge S 2016 *Intersectionality (Key Concepts)* (Cambridge: Polity Press)

[12] Rosa K, Blue J, Hyater-Adams S, Cochran G and Prescod-Weistein C 2021 Resource letter RP-1: race and physics *Am. J. Phys.* **89** 751–68

[13] Ong M 2023 *The Double Bind in Physics Education* (Cambridge, MA: Harvard Education Press)

[14] Atherton T J, Barthelemy R S, Deconinck W, Falk M L, Garmon S, Long E, Plisch M, Simmons E H and Reeves K 2016 *LGBT Climate in Physics: Building an Inclusive Community* (College Park, MD: American Physical Society)

[15] Cech E and Waidzunas T 2021 Systemic inequalities for LGBTQ professionals in STEM *Sci. Adv.* **7** eabeo933

[16] Cavanagh S R 2016 *The Spark of Learning: Energizing the College Classroom with the Science of Emotion* (Morgantown, WV: West Virginia University Press)

[17] Materials for teachers and students: teaching guides and educational games on history of the physical sciences *American Institute of Physics* https://aip.org/history-programs/physics-history/teaching-guides (Accessed: 13 June 2024)

[18] National Public Radio 2019 *Wait, Wait, Don't Tell Me* (podcast) 2 Nov https://www.npr.org/programs/wait-wait-dont-tell-me/775282313/wait-wait-for-nov-2-2019-with-not-my-job-guest-gloria-steinem

[19] Hyde J 2005 The gender similarities hypothesis *Am. Psychol.* **60** 581–92

Chapter 3

Supporting theories

In my own small experience, sexism has been something very simple: the cumulative weight of constantly being told that you can't possibly be what you are.

—Hope Jahren, *Lab Girl* [1]

As we will see in other chapters, the number of women in physics is nowhere near representative of women in the broader population. And it's been this way for as long as physics has been a field in its own right. (But someday that will change!) Trying to determine the causes of this under-representation has kept researchers busy for many decades.

There have been many theories proposed to explain women's lack of participation in physics. Some have been disproven, such as the idea that women are inherently, biologically less able at spatial reasoning. Some have been developed recently and are still being examined. In this chapter I share the basics of a few of the theories the literature is based on, so that readers have context for the research.

3.1 Unconscious bias

The last two decades have produced a great deal of research on the effects our unconscious biases may have on our actions. Implicit (unconscious, unexamined) bias is bias we are unaware of, typically a product of our society and upbringing. If you grow up in a culture that is sexist and racist, you will unknowingly absorb those biases. Unconscious bias has nothing to do with our professed, explicit beliefs. Thankfully, these biases are malleable. Two good books on this are *Blind Spot* [2] and *The End of Bias: A Beginning* [3].

While we may not know about our unconscious biases, they have an impact on what we do, how we perceive others, and how we feel. Bias against gender means that identical resumes, one with a male name and one with a female name, will produce fewer employment offers and a lower initial salary offer for the female [4].

doi:10.1088/978-0-7503-6435-5ch3

This may be exacerbated when the reviewer is a male of low power [5]. A CV study on physics professors [6] showed that faculty viewed male candidates as more competent and more hirable than identical female candidates. Asian and White candidates were preferred over Black and Latinx candidates as well. Though invisible, these biases affect us.

A fascinating series of implicit association tests (IATs) are available from the Project Implicit website (https://implicit.harvard.edu/implicit/). These tests have been shown to uncover the unknown, implicit biases that we all carry. One of these implicit biases is a connection between science and men. The unconscious connection between science and men, but not women, can affect not only a person's decision whether or not to go into science [7], but can also relate to achievement in science. Countries with a stronger correlation between men and science on the implicit bias test exhibit higher scores by boys on the TIMSS 8th grade science tests [8]. Higher national gender–science stereotypes also correlate with lower numbers of women in science [9]. Readers should note that the IAT results are quite variable even for a single individual. Of particular interest, exposure to counter-stereotypes before the test can change the results [10].

The Women In Science and Engineering Leadership Institute (WISELI) group in Wisconsin, USA, has an IAT for women and leadership, and you can participate in their research by taking the test at https://wiseli.wisc.edu/research/gender-leadership-iat/. The connections between women and leadership provide one lens for examining the issue of women's participation in science. If science is a high-status field [11], then scientists are viewed as leaders. If women aren't viewed as belonging in science, and they aren't viewed as leaders, they face a double challenge. As one young woman in high school says: 'I guess scientists, you can say, have power. I don't know. And a lot of people don't like the idea of women having power.' [12] WISELI has a brochure on 'Advancing women in science and engineering: advice to the top' [13] as well as many other helpful resources [14].

While there are plenty of data on implicit bias in the sciences generally, there is nearly nothing specifically on physics. Given the reputation of physics as a field where success is perceived to require brilliance [15], physics would make an interesting population to study more closely. A small sample of physics academics in Ireland showed gender differences in implicit bias: men were more likely than women to show a stronger correlation between male and science [16].

These implicit biases have been studied carefully, and we know that people who profess an unbiased view can still hold these hidden ideas [2]. So what can be done about them? Well, it is possible to reset these biases! Several studies have shown that long-term changes can be made to our hidden beliefs [17]. The first step is to become aware of our biases. Next, we need to realize the effects these biases have, and to really care about changing them. Then we need to confront our biases by exposing ourselves to counter-bias information. Pictures of female scientists, research on girls' abilities in science, stories and biographies of women in science, recognition of male failures in science—all of these can help us overcome biases we may have against women in science. (The American Institute of Physics has a resource collection for teaching about women and minorities in the physical sciences [18].) It can be done, but first and critically we need to recognize our biases and feel motivated to change. Awareness, motivation, action.

There are plenty of ideas on how to reduce bias, from different perspectives [3, 19] and with varying degrees of research to support their possible effectiveness. At the departmental level, interventions can help promote less biased views: one study found that a 2.5 hour workshop led to reduced bias in departments across a university [20]. But be cautious—diversity trainings overall have mixed results [21] and should be designed carefully, not a slapdash hour of displaying slides simply to check the 'offered diversity training' box.

3.2 Stereotype threat

Our brains are very good at taking in large amounts of information and sorting it, prioritizing it, analyzing it. Much of this happens extremely quickly, including making decisions based on that information [22]. It is an evolutionary advantage to be able to sort things quickly into groups: tiger (will try to eat me) or deer (good to eat). This quick sorting is a heuristic our brains use. Heuristics can be good or bad; stereotypes are one type of heuristic. 'A stereotype is a type of heuristic in which a person uses cultural information about a group, stored in memory, to evaluate an individual in ways that can be altogether hallucinatory [3].'

When a society has biases against particular groups, those biases often create stereotypes, such as 'girls can't do physics'. If a stereotyped person is reminded of those stereotypes, it can change their behavior and performance in the direction of the stereotype. This is called stereotype threat.

Claude Steele describes stereotype threat very well in his book *Whistling Vivaldi* [23]. In a seminal experiment, African American and White students were given a difficult test. One group was told that the test was a good measure of verbal ability; the other group was told it was simply a problem-solving exercise and did not diagnose ability. The first group's African American students performed more poorly than their White counterparts; the second group showed no significant difference. By being subtly reminded of the stereotype that African Americans are less good at verbal tests, students in the stereotyped group did less well than when they were not reminded of this stereotype.

This experiment has been conducted with women and math tests, with women in the threat condition (they were reminded of the stereotype) doing much worse than both men and women in a non-threat condition [24]. One article claims that by simply waiting to ask about gender until AFTER taking an AP test rather than posing such demographic questions beforehand, women do significantly better [25]. Even so small thing as writing down your name, major, and gender before a test can trigger a stereotype threat response, and reduce your performance. However, this effect may be lessening [26], as agreement with the stereotype diminishes in our culture [27].

In physics, one study [28] demonstrated that not only does a reminder of the stereotype reduce young women's performance on typical quantitative physics problems, explicitly stating the opposite ('No gender differences in performance have been found on this test.') reduced the gender gap to zero. They also compared explicit threat ('This test has shown gender differences with males outperforming females on the problems.') to implicit threat (no mention of gender or performance). Results between explicit and implicit threat conditions were the same, suggesting that awareness of the stereotype is all around us all the time. And another reminder that our environment affects us: feelings of stereotype threat are more severe in a (male-dominated) physics classroom than in a (female-dominated) biology classroom [29].

Other studies in physics note that test anxiety on a timed test may trigger stereotype threat [30], and (unfortunately) that active learning may have a negative effect on stereotype threat on women in an electricity and magnetism unit [31]. It is possible that this is due to women who agree with the stereotype experiencing stronger stereotype threat than women who do not agree with the stereotype [32].

Comparing women in biology to women in physics classes, stereotype threat might be reducing women's science identity (see section 3.4 below) by lowering the perceived value/usefulness of science. It may also lower women's science career intentions for women in physics [33].

Fortunately, just as with implicit bias, there are things we can do to reduce the effects of stereotype threat. One fascinating approach is employing self-affirmation, with a short exercise in which students are asked to write down values that are important to them and why. This exercise can lead to reduced effects from stereotype threat [34] for women and math tests. In physics, this has been successfully tested at the college level [35], partially replicated [36], and not replicated [37].

3.3 Sense of belonging

What makes one person decide to stick with physics, while another may give up? One of the better predictors for women in physics is whether or not they have a *sense of belonging*. At the basic level, the construct of 'a sense of belonging' is as simple as it reads: do you feel like you belong in a certain environment? As a more formal psychological construct, it probably started in 1995 with Baumeister and Leary's paper 'The need to belong: desire for interpersonal attachments as a fundamental human motivation [38]'.

In male-dominated STEM fields, women have a harder time feeling like they belong [39]. No surprise there. In physics, perhaps one of the earliest uses of the phrase is in 2006. In a study funded by the IOP, the phrase 'sense of belonging' was used more generally, but was consistent with the more modern understanding of the phrase [40].

Women in an undergraduate physics course have reported lower sense of belonging, and this is linked to persistence for women [41]. In another study, seeing a video of a gender-imbalanced STEM conference reduced women's sense of belonging but not men's [42]. Sense of belonging for women in physics class is also connected to how strongly women endorse the stereotype that women are less good at physics [43]. For women of color, the sense of belonging is typically even lower than for White women.

Sexual harassment is another factor that can affect a woman's sense that she belongs in physics. Among US women physics undergraduates, nearly three quarters of them had experienced some form of sexual harassment [44]. Having experienced *sexist* harassment ('Women just aren't that good at science' or being put down because of gender) was strongly tied to a negative sense of belonging.

For a nice overview of the issues relating to a sense of belonging for women in physics, read 'Fitting in or opting out: a review of key social-psychological factors influencing a sense of belonging for women in physics' [45].

3.4 A science identity and self-efficacy

A related concept to sense of belonging is the idea of science identity. Based on work by Carlone and Johnson [46], the idea of science identity is at heart the idea 'I am a science person'. Many factors feed into the building of science identity. You can imagine that it starts early; early home environment [47] and out-of-school science activities [48] can have effects on STEM/physics identity, and can help predict future STEM career choice.

Science identity comprises three parts: competence, performance, and recognition [46]. Recognition or acknowledgment is an important component of building a science or math identity [49]. It is hard to maintain 'I am a physics person' when no one around you says, 'You are a physics person'.

Competence and performance are parts of another concept called self-efficacy: the belief in your ability to complete a task [50]. Self-efficacy is field- and task-specific. Women in physics tend to have lower self-efficacy in physics [51], possibly affected by teaching styles [52].

Stereotype threat can also affect science identity: women in physics classes reported more concerns of stereotype threat than women in biology classes, and higher stereotype threat correlated with lower science identity [29].

Hazari and coauthors make a good argument that rather than looking at 'science identity', it is more valuable to look at specific fields: we want to know about people's 'physics identity' [53]. This is particularly important for gender distinctions and differences. Physics is male dominated; biology is not. Much of the work in physics identity has been done by teams including Hazari and researchers at Florida International University.

At this point the reader can guess that women are less likely to hold a physics identity than men. Yup [48, 53]—for so many reasons, a good number of which are listed above. This topic has been a popular one in the last 10–15 years, which is very helpful for the purposes of this book, and there is a great deal to read about it.

My reading of the literature here leads me to feel that the younger years (college and earlier) are definitely the most important to building one's physics identity. As one undergraduate said: 'I was aware that particularly physics is not a very women heavy field … it was not overwhelmingly encouraging. I think that is particularly more of a problem early on [54].'

3.5 Other concepts

Since issues of women and physics are cultural, there are many more theories and concepts that have been studied. I will list a few that readers may want to explore:

- Science anxiety
- Motivation: intrinsic and extrinsic
- Imposter phenomenon
- External/internal validation
- Feminist critiques of science/feminist philosophy of science

I hope that the reader will keep in mind these underlying concepts as they read further in the book. Rather than go into the details of, say, how tokenism may affect physics identity, I will simply refer to physics identity and readers can jump back here for context.

References

[1] Jahren H *Lab Girl* (New York: Knopf) pp 182–3
[2] Banaji M and Greenwald A 2013 *Blind Spot* (New York: Random House)
[3] Nordell J 2021 *The End of Bias: A Beginning* (New York: Metropolitan) p 164
[4] Moss-Racusin C, Dovidio J, Brescoll V, Graham M and Handelsman J 2012 Science faculty's subtle gender biases favor male students *Proc. Natl Acad. Sci.* **109** 16474–9
[5] Hoover A, Hack T, Garcia A, Goodfriend W and Habahi M 2019 Powerless men and agentic women: gender bias in hiring decisions *Sex Roles* **80** 667–80
[6] Eaton A, Suanders J, Jacobson R and West K 2020 How gender and race stereotypes impact the advancement of scholars in STEM: professors' biased evaluations of physics and biology post-doctoral candidates *Sex Roles* **82** 127–41
[7] Smyth F L, Greenwald A G and Nosek B A 2009 Implicit gender-science stereotype outperforms math scholastic aptitude in identifying science majors (unpublished)
[8] Nosek B A *et al* 2009 National differences in gender-science stereotypes predict national sex differences in science and math achievement *Proc. Natl Acad. Sci. USA* **106** 10593–7
[9] Miller D, Eagly A and Linn M 2015 Women's representation in science predicts national gender-science stereotypes: evidence from 66 nations *J. Educ. Psychol.* **107** 631–44
[10] Burns M D, Monteith M J and Parker L R 2017 Training away bias: the differential effects of counterstereotype training and self-regulation on stereotype activation and application *J. Exp. Soc. Psychol.* **73** 97–110
[11] Occupational Prestige Ratings https://occupational-prestige.github.io/opratings/opratings.html (Accessed: 20 June 2024)
[12] Grossman J and Porche M 2014 Perceived gender and racial/ethnic barriers to STEM success *Urban Educ.* **49** 698–727
[13] Handelsman J, Sheridan J, Fine E and Carnes M 2009 Advancing women in science and engineering: advice to the top *WISELI Brochure* https://wiseli.wisc.edu/wp-content/uploads/sites/662/2018/11/AdviceTopBrochure.pdf (Accessed: 13 June 2024)
[14] WISELI http://wiseli.wisc.edu/ (Accessed: 13 June 2024)
[15] Leslie S-J, Cimpian A, Meyer M and Freeland E 2015 Expectations of brilliance underlie gender distributions across academic disciplines *Science* **347** 262–5

[16] Poppenhaeger K 2019 Unconscious bias in academia: from PhD students to professors *AIP Conf. Proc.* **2109** 130001

[17] Devine P, Forscher P, Austin A and Cox W 2012 Long-term reduction in implicit race bias *J. Exp. Soc. Psychol.* **48** 1267–78
Lebrecht S, Pierce L J, Tarr M J and Tanaka J W 2009 Perceptual other-race training reduces implicit racial bias *PLoS ONE* **4** e4215
Dasgupta N and Greenwald A G 2001 On the malleability of automatic attitudes: combating automatic prejudice with images of admired and disliked individuals *J. Pers. Soc. Psychol.* **81** 800–14

[18] Materials for teachers and students: teaching guides and educational games on history of the physical sciences American Institute of Physics https://aip.org/history-programs/physics-history/teaching-guides (Accessed: 13 June 2024)

[19] Eberhardt J 2020 *Biased* (New York: Penguin)
Hoffman D and Winter H 2022 Follow the science: proven strategies for reducing unconscious bias *Harvard Neg. Law Rev.* **28** 1–63

[20] Carnes M *et al* 2015 The effect of an intervention to break the gender bias habit for faculty at one institution *Acad. Med.* **90** 221–23

[21] Bezrukova K, Spell C, Perry J and Jehn K 2016 A meta-analytical integration of over 40 years of research on diversity training evaluation *Psychol. Bull.* **142** 1227–74

[22] Gladwell M 2005 *Blink* (New York: Back Bay)

[23] Steele C 2010 *Whistling Vivaldi* (New York: Norton)

[24] Spencer S, Steele C and Quinn D 1999 Stereotype threat and women's math performance *J. Exp. Soc. Psychol.* **35** 4–28

[25] Danaher K and Crandall C 2008 Stereotype threat in applied settings re-examined *J. Appl. Soc. Psychol.* **38** 1639–55

[26] Karim N 2018 Impact of evidence-based active-engagement courses on student performance and gender gap in introductory physics *PhD Dissertation* University of Pittsburgh, Pittsburgh

[27] Maries A, Karim N and Singh C 2019 Does stereotype threat affect female students' performance in introductory physics? *AIP Conf. Proc.* **2109** 120001
Gutmann B and Steltzer T 2021 Values affirmation replication at the University of Illinois *Phys. Rev. Phys. Educ. Res.* **17** 020121
Lauer S, Momsen J, Offerdahl E, Kryjevskaia M, Christensen W and Montplaisir L 2013 Stereotyped: investigating gender in introductory science courses *CBE Life Sci.* **12** 30–8

[28] Marchand G and Taasoobshirazi G 2013 Stereotype threat and women's performance in physics *Int. J. Sci. Educ.* **35** 3050–61

[29] Smith J, Brown E, Thoman D and Deemer E 2015 Losing its expected communal value: how stereotype threat undermines women's identity as research scientists *Soc. Psychol. Educ.* **18** 443–66

[30] Burkholder E and Salehi S 2022 Exploring the pre-instruction gender gap in physics *PLoS One* **17** e0271184

[31] Maries A, Karim N and Singh C 2020 Active learning in an inequitable learning environment can increase the gender performance gap: the negative impact of stereotype threat *Phys. Teach.* **58** 430–3

[32] Maries A, Karim N and Singh C 2018 Is agreeing with a gender stereotype correlated with the performance of female students in introductory physics? *Phys. Rev. Phys. Educ. Rev.* **14** 020119

[33] Deemer E, Thoman D, Chase J and Smith J 2014 Feeling the threat: stereotype threat as a contextual barrier to women's science career choice intentions *J. Car. Dev.* **41** 141–59

[34] Martens A, Johns M, Greenberg J and Schimel J 2006 Combating stereotype threat: the effect of self-affirmation on women's intellectual performance *J. Exp. Soc. Psychol.* **42** 236–43

[35] Miyake A, Kost-Smith L E, Finkelstein N D, Pollock S J, Cohen G L and Ito T A 2010 Reducing the gender achievement gap in college science: a classroom study of values affirmation *Science* **330** 1234–7

[36] Kost-Smith L, Pollock S, Finkelstein N, Cohen G L, Ito T A and Miyake A 2012 Replicating a self-affirmation intervention to address gender differences: successes and challenges *AIP Conf. Proc.* **1413** 231
Lauer S, Momsen J, Offerdahl E, Kryjevskaia M, Christensen W and Montplaisir L 2013 Stereotyped: investigating gender in introductory science courses *CBE Life Sci.* **12** 30–8

[37] Gutmann B and Steltzer T 2021 Values affirmation replication at the University of Illinois *Phys. Rev. Phys. Educ. Rev.* **17** 020121

[38] Baumeister R F and Leary M R 1995 The need to belong: desire for interpersonal attachments as a fundamental human motivation *Psych. Bull.* **117** 497–529

[39] Seymour E and Hunter A B 2019 *Talking About Leaving Revisited* (Cham: Springer)

[40] Murphy P and Whitelegg E 2006 Girls and physics: continuing barriers to 'belonging' *Curric. J.* **17** 281–305

[41] Lewis K, Stout J, Finkelstein N, Pollock S, Miyake A, Cohen G and Ito T 2017 Fitting in to move forward: belonging, gender, and persistence in the physical science, technology, engineering, and mathematics (pSTEM) *Psychol. Women Quart.* **41** 420–36

[42] Murphy M, Steele C and Gross J 2007 Signaling threat: how situational cues affect women in math, science, and engineering settings *Psychol. Sci.* **18** 879–85

[43] Stout J, Ito T, Finkelstein N and Pollock S 2013 How a gender gap in belonging contributes to the gender gap in physics participation *AIP Conf. Proc.* **1513** 402–5

[44] Aycock L, Hazari Z, Brewe E, Clancy K, Hodapp T and Goertzen R M 2019 Sexual harassment reported by undergraduate female physicists *Phys. Rev. Phys. Educ. Rev.* **15** 010121

[45] Lewis K, Stout J, Pollock S, Finkelstein N and Ito T 2016 Fitting in or opting out: a review of key social-psychological factors influencing a sense of belonging for women in physics *Phys. Rev. Phys. Educ. Rev.* **12** 020110

[46] Carlone H and Johnson A 2007 Understanding the science experiences of successful women of color: science identity as an analytic lens *J. Res. Sci. Teach.* **44** 1187–218

[47] Dou R, Hazari Z, Dabney K, Sonnert G and Sadler P 2019 Early informal STEM experiences and STEM identity: the importance of talking science *Sci. Educ.* **103** 623–37

[48] Lock R, Hazari Z and Potvin G 2019 Impact of out-of-class science and engineering activities on physics identity and career intentions *Phys. Rev. Phys. Educ. Rev.* **15** 020137

[49] Godwin A, Potvin G, Hazari Z and Lock R 2015 Identity, critical agency, and engineering majors: an affective model for predicting engineering as a career choice *J. Eng. Educ.* **105** 312–40
Cribbs J D, Hazari Z, Sonnert G and Sadler P 2015 Establishing an explanatory model for mathematics identity *Child Dev.* **86** 1048–62

[50] Bandura A 1977 Self-efficacy: toward a unifying theory of behavioral change *Psychol. Rev.* **84** 191–215

[51] Shaw K 2003 The development of a physics self-efficacy instrument for use in the introductory classroom *AIP Conf. Proc.* **6** 020112

Cavallo A M L, Potter W H and Rozman M 2004 Gender differences in learning constructs, shifts in learning constructs, and their relationship to course achievement in a structured inquiry, year-long college physics course for life science majors *Sch. Sci. Math.* **104** 288–300

Fencl H and Scheel K 2006 Making sense of retention: an examination of undergraduate women's participation in physics courses *Removing Barriers: Women in Academic Sicence, Technology, Engineering, and Mathematics* ed J M Bystydzienski and S R Bird (Bloomington, IN: Indiana University Press) pp 287–302

[52] Fencl H and Scheel K 2005 Engaging students: an examination of the effects of teaching strategies on self-efficacy and course climate in a nonmajors physics course *J. Coll. Sci. Teach.* **35** 20–4

Sawtelle V 2011 A gender study investigating physics self-efficacy *PhD Dissertation* Florida International University, University Park, FL

[53] Hazari Z, Sadler P and Sonnert G 2013 Research and teaching. The science identity of college students: exploring the intersection of gender, race, and ethnicity *J. Coll. Sci. Teach.* **42** 82–91

[54] Eren E 2021 Exploring science identity development of women in physics and physical sciences in higher education: a case study from Ireland *Sci. Educ.* **30** 1131–58 1144

IOP Publishing

Women and Physics (Second Edition)

Laura McCullough

Chapter 4

How many women are in physics?

> Science is not a boy's game, it's not a girl's game. It's everyone's game.
> It's about where we are and where we're going.
> —Nichelle Nichols (Lt Uhura from *Star Trek*)

Across the globe, girls and young women get excited about physics. Yet as they move forward with their chosen field, they will encounter fewer and fewer 'fellow' women. In this chapter we are going to explore the representation of women in physics from secondary school through graduate education. How many women are studying physics and choosing physics as their career?

4.1 Historical women in physics

A good start is to take a brief look back through the years, to remind ourselves that women have been doing physics for centuries [1]. There are numerous biographies recounting the lives of women in physics. The most famous female physicist is, of course, Marie Curie. But if you ask someone to name a second female physicist, they are likely to draw a blank. Yet women in physics and astronomy have made many significant contributions to the field. Increasing awareness of the work that women have done in physics is a worthwhile goal, and many websites are working to address this [2].

The issues that women in physics faced up through the early twentieth century were quite different than what women face today. It was a very difficult time. They weren't allowed in universities, tutors would often refuse to teach women, and women had no right to own property until the nineteenth century in many countries. (Read Fara's *A Lab of Her Own* for some amazing stories! [3]) So how did women succeed? Usually by working with a male scientist.

Doing formal physics research typically requires access to journals, equipment, and lab space. Supportive males, often a parent, brother, or spouse [4], allowed women access to the labs and equipment that they needed to do their physics. Marie

doi:10.1088/978-0-7503-6435-5ch4

Curie worked with her husband and her daughter. Caroline Herschel worked alongside her brother, Sir William Herschel. Mileva Einstein-Marić was a physicist in her own right before she married Albert Einstein. Lise Meitner worked as an unpaid assistant before collaborating with Otto Hahn. Having a supportive husband has helped several famous physicists, and yet it stood in the way sometimes: anti-nepotism laws prevented Nobel Prize winner Maria Goeppert Mayer from being paid to work at the same university at which her husband worked [5].

Because women working in physics were often treated as unpaid assistants, few ever received recognition for their work. It wasn't until women were allowed to enter universities and scientific societies in their own right that they began to be generally credited for their work.

There are untold numbers of women who have loved physics and contributed to the field, yet many of them are lost to history because they were never acknowledged in their time. You can thank Margaret Rossiter for bringing many of them back into the light [6]. One place where progress has been made is that it is now expected that anyone who works on a project will be credited for their work. Journals and professional organizations now have policies and ethics codes explicitly demanding this [7]. This helps us develop a picture of how many women are working in physics currently. By looking at national data sources, we can now get a strong sense of the participation of women in physics today. Let's start by looking at the earliest formal physics instruction, which usually happens in high school/upper secondary education.

4.2 When science was for women

Most people likely believe that science education has always been a male domain. Yet there was a period of time where science was viewed as something to teach young women, while young men needed to learn the more important classical education of Greek, Latin, mathematics, history, and rhetoric.

During the last half of the nineteenth century, schools across America and the United Kingdom were focusing the education of young women on science, typically considered 'domestic science' or 'household science'. Since women were the home-makers and in charge of the health and cleanliness of the home, they were encouraged to learn the basics of science [8]. Science was also purported to develop strong thinking skills for women [9], in the same way that learning the classics did for men.

Ironically, at the same time, women were discouraged from higher education because of fears that such education would create 'pale, weak, neuralgic, dyspeptic, hysterical, menorraghic, dysmenorrhoeic girls and women' [10]. Much medical research at the time focused on proving that education was harmful to women, particularly that the brain will pull essential energy and nutrients away from the uterus [11]. (For a history of how science has been used to keep women as the 'lesser' sex, try reading Saini's *Inferior* [12].) One writer in 1735 argued *for* the education of women, including this rationale: 'If the female Tongue will be in Motion, why should it not be set to go right? Could they discourse about the Spots in the Sun, it might divert them from publishing the Faults of their Neighbours ...' [13] Science can prevent gossiping? If only that author could see scientists' social media today.

Physics education as a discipline was slowly developing at the same time. By the late nineteenth century, educators were worried about the lack of standardization of education in the US, and the 'Committee of Ten' was formed in 1892 [14]. This committee pulled together science educators and scientists, who recommended the now-common order of high school science as biology, then chemistry, then physics. The committee itself recommended chemistry after physics, but the scientists' recommendation was more generally accepted. And thus began more than a century of physics being last in the high school science curriculum [15].

4.3 Current trends in women's participation in physics

The slow drain of women leaving physics and other sciences as they move into higher levels of the field, or—too often—don't, has often been referred to as the leaky pipeline problem. The analogy works well in some ways, with the initial talent pool of physicists being the initial flow of water into a pipe. But along the way, at each of the major junctures (high school to college, college to graduate school, etc) the field loses women at a higher rate than it loses men. The pipeline analogy has its problems [16], and I support the more appropriate analogy of a physics 'highway' [17]. There are off-ramps aplenty, but there are also on-ramps. Some go fast on the highway, some move slowly.

Whereas we can look at many of the problems women encounter in physics in a more qualitative way, the issue of falling participation at higher levels in the Anglophone countries is best approached by looking at the statistics. So, we're going to do just that, following the data as it moves upward, starting with the educational levels where most students first encounter physics as a standalone topic.

4.4 Secondary education

In the US most schools do not have a separate physics class until high school.

Most other Anglophone countries are similar. Science is considered an important subject to take as preparation for college (tertiary education), so most young men and women in the US are encouraged to take a physics class in high school though most are not required to do so. In the UK and Australia, physics is optional once students are 16 years old.

In the US, the number of high school students taking physics is 42%, a very respectable number [18]. The number of women in high school physics classes as a percentage of the total is about 47% and has stayed consistent over the last decade— near parity [19]. But girls are less likely to take the higher levels of high school physics (they are 25%–40% of Advanced Placement students) [20], and are less likely to pass the AP exams [21].

We find less encouraging numbers in the UK. In England 2.6% of girls choose to take A-level physics and 8.6% of boys [22]. Scotland shows approximately one-quarter women among National level passes in physics [23] and 27.5% of girls choosing Higher Physics compared to 72.5% of boys; this gives about a 30% rate of girls in upper-level physics. In Wales only 2% of girls choose A-level physics, and 9% of boys. In Northern Ireland, only 1% of girls choose A-level, and only 2.6% of boys. Interestingly, single-sex schools do a better job of sending girls on to physics [24]. Overall, in the UK, A-level courses have 23% girls [25]. In Ireland, the number of girls with a physics Leaving Certificate has held constant at 14% for several years [26].

New Zealand is in line with the US for participation rates, with girls making up 40% of the population of the average physics classes in Year 12 or 13 [27]. Australia's percentages are closer to the UK, with about 22% of girls choosing physical science in Year 12 compared with 36% for boys [28]. Girls were about one-quarter of Year 12 physics and astronomy enrollments in 2021 [29].

Canada also has fewer girls in high school physics across most of the provinces, where they are participating at about half the number of boys. The exceptions to this rule are Quebec and Saskatchewan which have had near equal numbers of boys and girls in physics [30]. A study in Ontario shows the number of girls taking grade 11 and 12 physics at 40% and 34% respectively; this hasn't changed much recently [31].

With the ratio so variable across countries and even provinces it can be hard to make exact statements about the magnitude of the problem, but it is clear that work needs to be done at the pre-high school level if we want to see parity in students beginning formal education in physics.

4.5 College/tertiary education

Students often have very limited choices in which courses they will take in high school, so it is not until they enter university that we see them taking full charge of their educational destiny. Unfortunately, one of the choices that young women often make at this point is to move away from physics. This is where we see our next significant drop in the number of young women choosing to pursue physics as they opt out of taking physics classes and majoring in physics.

In the US only one in five physics bachelor's degrees go to a woman [32]. There has been a very slow growth in that number over the last forty years, since only about 10% of bachelor's degrees went to women in 1981, but at that rate parity is a long way in the future. For more than 20 years, the National Science Foundation has issued regular reports on 'Women, minorities, and persons with disabilities', which is a nice way to look at historical trends [33].

One analysis offers up estimates of when various fields will reach gender parity, using publication data from more than 100 countries [34]. Physics remains at the low end, with their estimate of gender equality taking more than 100 years. Only computer science and quantitative finance are worse, among over 50 fields. Computer science is an example of rapid cultural change: the field was dominated by women from the 1960s through the mid-1980s, and since then the percentage of women in the field has been declining slowly [35].

Interestingly, despite the significant difference in participation at the high school levels the current US number of approximately 20% is quite similar to the UK, where women make up around 23% of the undergraduate physics population [25].

In Aotearoa/New Zealand, the numbers of physics bachelor's degrees tend to be small, with fewer than 150 a year. In 2022, physics bachelors were awarded to 40 women (33%), 75 men, and 5 of other gender. The Bachelor's Honors numbers included 15 women (30%) and 35 men [36].

Other Anglophone countries also show under-representation in women's participation in physics at the undergraduate level. In Ireland the 2019 undergraduate participation rate was 28% [26]. Australian university physics enrollments are at 25% women [37].

A preprint in 2024 describes the first large study aimed at diversity and equity in Canadian physics [38]. They estimate approximately 3000 undergraduate students (not degrees conferred, but active students), and in the survey sample, 46% of undergraduates self-reported as women or gender diverse. Wow! The authors do note that a high proportion of students responded to the survey, younger generations are more likely to disclose gender identity and younger generations are more interested in diversity and equity issues. A better estimate may come from Ontario, where women make up 20% of physics classes and degrees [39].

4.6 Postgraduate education

For women in the US, the percentage receiving master's degrees in physics runs at about 24% and then drops back to 21% for doctoral degrees going to women (2017–2021 average) [32, 40]. This is following the general upward trend we see in bachelor's degrees, with a steady but slow climb over time in the participation of women in graduate physics.

Canadian enrollment for math and physical sciences shows about 35% women at the master's level, and 30% for doctorates. Though they do not break out the numbers for the two fields individually we can look at the general pattern of higher percentages of women in math, and infer that the physics numbers are likely significantly less than 30% [9]. The large survey of the CAP mentioned above estimates 500 master's and 500 doctoral students [38]. Among survey respondents, masters' students included 45% women and gender diverse people, and doctoral students were at 39%. However, the same caveat applies as with the undergraduate numbers.

While we see relatively static numbers in the US between degree levels, in the UK we see a small rise in the proportion of women in physics at higher degree levels with postgraduate physics participation rising to 26.5% from the 21.7% at the undergraduate level [41]. It is not clear what causes this increase, and more research in this area might prove very fruitful in service of the goal of promoting similar increases elsewhere in the English-speaking world.

4.7 Careers

Data for physics careers can be hard to find, especially among industry and governmental sectors. Academia is better studied (no surprise there), so we can obtain only a partial picture overall of how many women are actively involved in physics employment.

The Institute of Physics membership is 22% female and 1% gender diverse for those in employment in the UK [22]. Among higher education there are a few more women, at a whopping 26%. Data for Scotland are coalesced into STEM fields, where males (56%) outnumber females (44%) [23].

Australian data are unavailable, though there is a push to obtain country-wide data [42].

Higher education in the US shows 20% of physics faculty members are women, about the same rate as PhD earners [43]. A study of US physics PhD holders has a breakdown of where they were employed. Here are the percentages of women by area: academia (17%); government (15%); industry (9%); and other (16%) [44].

The best look at women in physics careers comes from a global survey. In countries with very high development indices the breakdown of women in various employment sectors shows some remarkable ratios. Academia (36%); government (44%); nonprofit (40%); industry (28%); self-employed (30%); primary/secondary school (48%) [45]. This suggests some self-selection into certain sectors of those with degrees in physics.

4.8 Conclusion

In this chapter, we have shown that women participate in physics at much lower rates than men do in English-speaking countries, though the numbers are slowly (very slowly) climbing. For over 50 years, scholars have asked what the reason for this lower participation is, and many ideas have been put forth to explain the discrepancy. The rest of this book is focused on different pieces of this puzzle. For a nice overview of research from 2019, try the 'Resource letter on gender and physics' [46].

References

[1] Zielinski S 2011 Ten historic female scientists you should know *Smithsonian Magazine* 19 Sept https://www.smithsonianmag.com/science-nature/ten-historic-female-scientists-you-should-know-84028788/ (Accessed: 13 June 2024)

[2] Some examples: Contributions of 20th century women to physics *CWP* http://cwp.library.ucla.edu/ (Accessed: 13 June 2024)
McKinnon M 2015 These 17 wome changed the face of physics *Gizmodo* https://gizmodo.com/these-17-women-changed-the-face-of-physics-1689043918 (Accessed: 13 June 2024)
Women in physics *Science Museum* https://sciencemuseum.org.uk/objects-and-stories/women-physics (Accessed: 13 June 2024)

[3] Fara P 2018 *A Lab of One's Own* (Oxford: Oxford University Press)

[4] Pycior H, Slack N and Abir-Am P (ed) 1996 *Creative Couples in the Sciences* (New Brunswick, NJ: Rutgers University Press)

[5] Struggle for employment: anti-nepotism laws in the academy *AIP Center for History of Physics* https://aip.org/sites/default/files/Struggle%20for%20Employment_Biographies%20Handout.pdf (Accessed: 13 June 2024)

[6] Rossiter M 1992 and 1995 *Women Scientists in America* (Baltimore, MD: Johns Hopkins University Press)

[7] For example, see Ethics guidelines *American Physical Society* https://aps.org/about/governance/policies-procedures/ethics (Accessed: 28 June 2024)

[8] Tolley K 2002 *The Science Education of American Girls: A Historical Perspective* (London: Routledge)

[9] Tolley K 1996 Science for ladies, classics for gentlemen *Hist. Educ. Quart.* **36** 129–53

[10] Clarke E 1873 *Sex in Education; or, a Fair Chance for Girls* (Boston, MA: Houghton Mifflin) p 62

[11] A fun modern take on this is in Ehrenreich B and English D 2005 *For Her Own Good: Two Centuries of the Experts' Advice to Women* (New York: Anchor)

[12] Saini A 2017 *Inferior* (Boston, MA: Beacon Press)

[13] Author unknown 1735 Arguments for educating women *New-York Weekly Journal* 19 May www.gilderlehrman.org/history-by-era/thirteen-colonies/resources/arguments-for-educating-women-1735 (Accessed: 13 June 2024)

[14] National Education Association of the United States. Committee of Ten on Secondary School Studies 1894 *Report of the Committee of Ten on Secondary School Studies: With the Reports of the Conferences Arranged by the Committee* (Cincinatti, OH: American Book Company)

[15] Sheppard K and Robbins D 2002 Lessons from the committee of ten *Phys. Teach.* **40** 426–31

[16] Xie Y and Shauman K 2003 *Women in Science: Career Processes and Outcomes* (Boston, MA: Harvard University Press)
Hammonds E and Subramaniam B 2003 A conversation on feminist science studies *Signs* **28** 923–44

[17] Anderson-Rowland M 2011 The engineering highway: a new metaphor especially appropriate for women *WEPAN 2009 National Conf. Proc.* https://journals.psu.edu/wepan/article/view/58580 (Accessed: 11 June 2024)

[18] Chu R Y and White S 2021 High school physics overview *AIP Report* https://ww2.aip.org/statistics/high-school-physics-overview (Accessed: 11 June 2024)

[19] Porter A M and Ivie R 2019 Women in physics and astronomy, 2019 *AIP Statistics Report* https://ww2.aip.org/statistics/women-in-physics-and-astronomy-2019 (Accessed: 11 June 2024)

[20] AIP 2019 Percent and number of young women who participated in AP physics exam, 2017 *AIP Data Graphic* https://ww2.aip.org/statistics/percent-and-number-of-young-women-who-participated-in-ap-physics-exam-2017 (Accessed: 11 June 2024)

[21] AIP 2019 Percent of young women and men who passed AP physics exams, 2017 *AIP Data Graphic* https://ww2.aip.org/statistics/percent-of-young-women-and-men-who-passed-ap-physics-exams-2017 (Accessed: 11 June 2024)

[22] Institute of Physics 2022 Written evidence submitted by Institute of Physics (Unpublished) *Unpublished by the Institute of Physics (IOP)* FSE0006 https://committees.parliament.uk/writtenevidence/107125/pdf/ (Accessed: 11 June 2024)

[23] Ekosgen 2017 Developing a Scottish STEM evidence base *Final Report for Skills Development Scotland* https://gov.scot/binaries/content/documents/govscot/publications/strategy-plan/2017/10/science-technology-engineering-mathematics-education-training-strategy-scotland/documents/00526538-pdf/00526538-pdf/govscot%3Adocument (Accessed: 11 June 2024)

[24] Institute of Physics 2018 Why not physics? A snapshot of girls' uptake at A-level *Report* https://iop.org/sites/default/files/2018-10/why-not-physics.pdf (Accessed: 13 June 2024)

[25] Jordan S, Bakewell S, Sadler W, Thiemann H, Wardlow J and Welsch C 2023 Women in physics in the United Kingdom: successes, challenges, and wider diversity *AIP Conf. Proc.* **3040** 050041

[26] Kavanagh Y, McLoughin E, Byrne M, Felton S, Walshe G and Whelan E 2023 Women in physics in Ireland—making the invisible visible *AIP Conf. Proc.* **3040** 050020

[27] Association for Women in the Sciences 2011 Women in science: a 2011 snapshot *AWIS* www.awis.org.nz/assets/Files/AWIS-Stats-2011-Booklet.pdf (Accessed: 13 June 2024)

[28] Sikora J 2014 *Gendered Pathways into The Post-Secondary Study of Science* (Adelaide: NCVER)
Science in Public 2014 Women in physics still going backwards *Report* www.scienceinpublic.com.au/media-releases/women-in-physics-aipc2014 (Accessed: 13 June 2024)

[29] Commonwealth of Australia 2024 STEM equity monitor *Data Report* https://industry.gov.au/publications/stem-equity-monitor (Accessed: 17 June 2024)

[30] NSERC 2010 Women in science and engineering in Canada *Report* Corporate Planning and Policy Directorate Natural Sciences and Engineering Research Council of Canada Ottawa, Ontario, Canada www.nserc-crsng.gc.ca/_doc/Reports-Rapports/Women_Science_Engineering_e.pdf (Accessed: 13 June 2024)
Predoi-Cross A, Austin R, Dasgupta A, Ghose S, Milner-Bolotin M, Steinitz M and Xu L 2013 Women in physics in Canada *AIP Conf. Proc.* **1517** 86–8

[32] Tyler J 2024 Degrees earned in the physical sciences and engineering fields *AIP Data Graphic* https://ww2.aip.org/statistics/physics-engineering-degrees-earned (Accessed: 11 June 2024)

[31] Wells M, Williams M, Corrigan E and Davidson V 2018 Closing the gender gap in engineering and physics *White Paper* http://onwie.ca/wp-content/uploads/2019/02/White-Paper-Final-Draft.pdf (Accessed: 11 June 2024)
Corrigan E, Williams M and Wells M 2023 High school enrolment choices—understanding the STEM gender gap *Can. J. Sci. Math. Techn. Educ.* **23** 403–21

[33] Document library US National Science Foundation https://nsf.gov/publications

[34] Holman L, Stuart-Fox D and Hauser C E 2018 The gender gap in science: how long until women are equally represented? *PLoS Biol.* **16** e2004956

[35] NCSES 2023 Diversity and STEM: women, minorities and persons with disabilities *Featured Report* https://ncses.nsf.gov/pubs/nsf23315/

[36] Tertiary achievement and attainment Education Counts https://educationcounts.govt.nz/statistics/achievement-and-attainment (Accessed: 10 June 2024)

[37] Maassoumi F, Maddison S and Foley C 2023 Women in physics in Australia 2021 *AIP Conf. Proc.* **3040** 050003
Fisher C, Brookes R and Thompson C 2022 'I don't study physics anymore': a cross-institutional Australian study on factors impacting the persistence of undergraduate science students *Res. Sci. Educ.* **52** 1565–81

[38] Hennessey E, Smolina A, Hennessey S, Tassone A, Jay A, Ghose S and Hewitt K 2024 Canadian physics counts: an exploration of the diverse identities of physics students and professionals in Canada arXiv: 2403.04670

[39] Wells M, Williams M, Corrigan E and Davidson V 2018 Closing the gender gap in engineering and physics *White Paper* http://onwie.ca/wp-content/uploads/2019/02/White-Paper-Final-Draft.pdf (Accessed: 11 June 2024)

[40] Diversity in physics *American Institute of Physics* https://aps.org/learning-center/statistics/diversity (Accessed: 11 June 2024)

[41] Institute of Physics 2010 Diversity in university physics: statistical digest 2010 *Report* https://warwick.ac.uk/fac/sci/physics/staff/working/juno/documentation/iop_diversity_in_university_physics.pdf (Accessed: 28 June 2024)

[42] Discussions at the *7th Int. Conf. on Women in Physics 2021*, Personal communications, 16 June 2024

[43] AIP 2023 Percent of physics faculty members who are women, 2002–2022 *AIP Data Graphic* https://ww2.aip.org/statistics/percent-of-physics-faculty-members-who-are-women-2002-2022 (Accessed: 11 June 2024)

[44] Czujko R and Anderson G 2015 Common careers of physicists in the private sector *AIP Report* https://aip.org/sites/default/files/statistics/phd-plus-10/PhysPrivSect.pdf (Accessed: 12 June 2024)

[45] Roy M F, Guillope C, Cesa M, Ivie r, White S, Mihaljevic H, Santamaria L *et al* 2020 A global approach to the gender gap in mathematical, computing, and natural sciences: how to measure it, how to reduce it *Zenodo*

[46] Blue J, Traxler A and Cochrane G 2019 Resource Letter: GP-1: gender and physics *Am. J. Phys.* **87** 616–26

Chapter 5

Home and family life

[O]nly a quarter of the three-year-old girls thought that their mother would want them to play with a baseball and mitt, or a skateboard …
—Cordelia Fine, *Delusions of Gender* [1]

We have seen that the proportion of women in physics in the English-speaking world is much lower than 50% at nearly every stage of education and career. What factors are causing this? Research on women and science has been conducted for over 50 years, and despite all we have learned, we are still discovering important new factors that play into this under-representation.

One obvious place to start examining any primarily cultural issue is with the family. Among women who choose physics, having the support of her family is an often-cited factor in going into physics [2]. Family support in general is important for all adolescents considering science, whether in the UK [3] or the US [4]. In one study of US undergraduates, 'the majority of students described parental encouragement and support for their choice of a STEM degree' [5]. The American Institute of Physics asked recent undergraduates what their influences were, and about 30% of both men and women named parents, relatives, or friends [6].

Even a single family member (or close family friend) who is an advocate for a young woman can be a major help to her. This support does not need to come from family in the sciences [7]; any family support will help with the motivation to study and work in science [3, 8]. And it doesn't need to be a mom either: both fathers and mothers can be influencers, though the higher the education of the parent the more likely they were to be called an influencer [4]. Interestingly, among US Latinos, parents are a key STEM influencer regardless of their level of education [9].

Successful physics girls may be more likely to have parents or a household rich in science—high science capital [10], although of course this is no guarantee for physics participation or science identity [11]. Another key factor may lie in the realm of science identity—having a parent who views you as 'a physics person'. Women in undergraduate physics were found to be less likely than men to believe that parents and others view them as a physics person [12].

'Twenty-one percent of girls said their parents encouraged them to be actresses and only 10 percent were encouraged to be engineers' [13], writes Karen Purcell. I find this to be an interesting data point, and highly quotable, but I also think it provides a good illustration of how carefully one needs to look at sources, both for context and methodology. For example, if 50% of girls expressed an interest in being an actress, and only 10% said they wanted to be engineers—a circumstance well within the realm of possibility—then another way of expressing the same data would be to say that only 50% of girls' parents supported their acting ambitions whereas 100% supported girls' desire to go into engineering. I originally considered removing the quote for this reason, but decided it would be more useful as an example of how important it is both to properly frame and carefully think about any research topic. I also continue to believe it is a useful framing tool to consider the way family interaction can impact the subject of science identity.

Without encouragement, and without ever being exposed to scientists, how are young girls and boys to learn that physics is a fun, satisfying, and productive career choice? 'Young girls cannot possibly consider opportunities they do not know exist' [14]. Teachers, parents and family members should show children that many science fields exist and are possibilities for *every* child.

We know that stereotypes about who is 'really, really smart' show up appallingly early; in the US six-year-old girls already believe that girls are less smart than boys [15]. And 10/11-year-old boys in the UK already show gendered views of science: '… don't think girls would make good scientists … because they aren't usually interested in science …' and '… most scientists wear glasses and girls these days care about fashion, and glasses aren't in fashion.' [16] While girls and boys may say that 'girly girls' could do physics, they still thought girly girls were less likely to choose smart or *hard* topics [17].

Early access to science (books, games, etc) as well as parent participation in science activities with a preschool child has been shown to improve science literacy for the next four years [18]. For older schoolchildren, having parents who enjoy physics and value physics as a career have strong associations with the child's future participation in physics [19]. In the same study, other factors influencing the likelihood of taking A-level physics include being a boy, parents thinking their child is academic, and parents thinking the child is interested in a male-dominated STEM field. Girls who were interested in future physics were perceived by their parents as being courageous and *not* being nurturing. (Sigh.) The authors note that this matches previous research that girls need to be 'exceptional' [quotation marks in original] in order to overcome the barriers to future physics participation.

An Australian study found that having lower socio-economic status (SES) parents was associated with lower interest in physics, and the effect was stronger for girls than boys [20]. This finding was supported by another study that found gender, indigeneity, and SES were predictors for lowered physics uptake [21]. An older study of New Zealand fourth and eighth grade students showed that SES status favored fourth grade boys' TIMSS science score, but the effect was gone for eighth graders [22]. Related to family and SES is video gaming behavior: a UK study found that male and female gamers were more likely to get a physical science degree, and female physical science majors were more likely to be heavy gamers, although SES influenced behavior as well [23].

Alexander *et al* [24] report that even at ages 4–7 there are gender differences in science interest and how families respond to their children's interest in science. And the source of science encouragement can be different for girls and boys: one study notes that females were more likely to be encouraged by teachers than parents, while males were encouraged by both approximately equally [25]. Also, 15-year-old girls in the UK reported less encouragement to go into physics from family and teachers than their male peers [26]. Intersectionality has also been looked at with mixed results: Caucasian and Latina girls were less likely to have parental support in physics than boys of both races [7], and Latina girls had the lowest personal beliefs in physics ability.

American Black girls may have stronger connections to parental support in school and science [27], which can provide some positive attitudes towards science, even though it is not realized in further participation [28]. Prior negative experience with systemic racism in their own education can influence how Black parents interact with their daughters [29]. Overall, however, this topic is poorly understood currently.

As is common with Black Americans, Latinx students in the US bring a stronger connection with family than their non-Hispanic peers, although parental engagement with their kids' schooling is different than 'traditional' (meaning White) parental involvement [30]. The use of 'traditional' as a stand-in for White, straight, cis, nuclear, etc, in this study is a good example of the inherently problematic nature of 'gap' or 'deficit' research as noted in chapter 3. I include it primarily because of the general lack of research in this area. We know very little about the interaction of Hispanic ethnicity with girls and science or physics.

Moving beyond the specifics of ethnicity and race, one large study found that parental pressure for STEM careers was more common in immigrant families [5].

Many female physicists report becoming interested in a physics career before they graduated from high school [31], demonstrating that encouragement during the teenage years can be very important to future career paths. The positive impacts on girls of supportive parents in regards to STEM careers lies on one side of a balance. On the other, we find that parents often hold beliefs that physics is a more masculine field [32].

One issue to consider is the line between support and pressure. Students who feel pressure from parents to enter STEM fields are more likely to leave their STEM major [5].

Family can be an important factor in girls' decisions to take physics courses and choose a physics career. However, we need to make sure that parents understand the value of (a) a physics career, and (b) encouraging their daughters to explore physics.

References

[1] Fine C 2010 *Delusions of Gender* (New York: Norton) p 202

[2] Taube J 2023 Understanding the experiences of young women in high school physics *EdD Dissertation* Southern New Hampshire University, Manchester and Hooksett, NH

[3] Mujtaba T and Reiss M 2014 A survey of psychological, motivational, family and perceptions of physics education factors that explain 15-year-old students' aspirations to study physics in post-compulsory English schools *Int. J. Sci. Educ.* **12** 371–93

[4] Sonnert G 2009 Parents who influence their children to become scientists: effects of gender and parental education *Soc. Stud. Sci.* **39** 927–41

[5] Seymour E and Hunter A B 2019 *Talking About Leaving Revisited* (Cham: Springer) p 381

[6] Mulvey P and Told J 2024 Physics bachelors: influences and backgrounds *AIP Report* https://ww2.aip.org/statistics/physics-bachelors-influences-and-backgrounds (Accessed: 11 June 2024)

[7] Simpkins S, Price C and Garcia K 2015 Parental support and high school students' motivation in biology, chemistry, and physics: understanding differences among Latino and Caucasian boys and girls *J. Res. Sci. Teach.* **52** 1386–407

[8] Aschbacher P, Li E and Roth E 2010 Is science me? High school students' identities, participation and aspirations in science, engineering, and medicine *J. Res. Sci. Teach.* **47** 564–82
Girl Scouts 2012 Generation STEM: what girls say about science, technology, engineering, and math *Report* www.girlscouts.org/content/dam/girlscouts-gsusa/forms-and-documents/about-girl-scouts/research/generation_stem_full_report.pdf (Accessed: 28 June 2024)
Archer L, Moote J, Francis B, DeWitt J and Yeomans L 2017 The 'exceptional' physics girl: a sociological analysis of multimethod data from young women aged 10–16 to explore gendered patterns of post-16 participation *Am. Educ. Res. J.* **54** 88–126

[9] Taningco M TTomas Riviera Policy Institute 2008 Latinos in STEM professions: understanding challenges and opportunities for next steps *Report* Tomas Riviera Policy Institute https://files.eric.ed.gov/fulltext/ED502064.pdf (Accessed: 28 July 2024)

[10] Archer L, Moote J, Francis B, DeWitt J and Yeomans L 2017 The 'exceptional' physics girl: a sociological analysis of multimethod data from young women aged 10–16 to explore gendered patterns of post-16 participation *Am. Educ. Res. J.* **54** 88–126
Archer L, Dawson E, DeWitt J, Seakins A and Wong B 2015 'Science capital': a conceptual, methodological, and empirical argument for extending Bourdieusian notions of capital beyond the arts *J. Res. Sci. Teach.* **52** 922–48

[11] Vedder-Weiss D 2018 'Won't you give up your snack for the sake of science?' Emerging science identities in family everyday interaction *J. Res. Sci. Teach.* **55** 1211–35

[12] Kalendar Z Y, Marshman E, Schunn C, Nokes-Malach T and Singh C 2019 Why female science, technology, engineering, and mathematics majors do not identify with physics: they do not think others see them that way *Phys. Rev. Phys. Edu. Rev.* **15** 020148

[13] Purcell K 2012 *Unlocking Your Brilliance: Smart Strategies for Women to Thrive in Science, Technology, Engineering and Math* (Austin, TX: Greenleaf) quoted in Sierra Nevada Journeys 2015 Girls in STEM www.sierranevadajourneys.org/snjblog/girlsinstem (Accessed: 28 June 2024)

[14] Purcell K 2012 An interview with Karen Purcell, advocate for women in STEM *Techniques* **87** 14

[15] Bian L, Leslie S-J and Cimpian A 2017 Gender stereotypes about intellectual ability emerge early and influence children's interests *Science* **355** 389–91

[16] Archer L, DeWitt J, Osborne J, Dillon J, Willis B and Wong B 2010 'Doing' science versus 'being' a scientist: examining 10/11-year-old schoolchildren's constructions of science through the lens of identity *Sci. Educ.* **94** 635

[17] Francis B, Archer L, Moore J, de Witt J and Yeomans L 2017 Femininity, science, and the denigration of the girly girl *Br. J. Soc. Educ.* **38** 1097–110

[18] Bae J, Shavlik M, Shatrowsky C E, Haden C A and Booth A E 2023 Predicting grade school scientific literacy from aspects of the early home science environment *Front. Psychol.* **14** 1113196

[19] Jones K and Hamer J 2022 Examining the relationship between parent/carer's attitudes, beliefs, and their child's future participation in physics *Int. J. Sci. Educ.* **44** 201–22

[20] Justman M and Mendez S 2018 Gendered choices of STEM subjects for matriculation are not driven by prior differences in mathematical achievement *Econ. Educ. Rev.* **64** 282–97

[21] Cooper G and Berry A 2020 Demographic predictors of senior secondary participation in biology, physics, chemistry, and earth/space sciences: students' access to cultural, social and science capital *Int. J. Sci. Educ.* **42** 151–66

[22] Chamberlain M 2003 The relative performance of girls and boys in science at grades 4 and 8 in New Zealand in 1994 and 1998 *Stud. Educ. Eval.* **29** 259–77

[23] Hosein A 2019 Girls' video gaming behaviour and undergraduate degree selection: a secondary data analysis approach *Comput. Hum. Behav.* **91** 226–35

[24] Alexander J, Johnson K and Kelley K 2012 Longitudinal analysis of the relations between opportunities to learn about science and the development of interests related to science *Sci. Educ.* **96** 763–86

[25] Dingel M 2006 Gendered experiences in the science classroom *Removing Barriers* ed J M Bystydzienski and S R Bird (Bloomington, IN: Indiana University Press)

[26] Mujtaba T and Reiss M 2013 What sort of girl wants to study physics after the age of 16? Findings from a large-scale UK survey *Int. J. Sci. Educ.* **35** 2979–98

[27] West-Olaunji C, Pringle R, Adams T, Baratelli A, Goodman R and Maxis S 2008 How African American middle school girls position themselves as mathematics and science learners *Int. J. Sci. Educ.* **14** 219–27

[28] Young J, Feille K and Young J 2017 Black girls as learners and doers of science: a single-group summary of elementary science achievement *Electron. J. Sci. Educ.* **21** 1–20
Russell M 2005 Untapped talent and unlimited potential: African American students and the science pipeline *Negro Educ. Rev.* **56** 167–82

[29] Consalvi D A 2023 Phenomenological exploration of how African American parents' lived experiences influence children's STEM choices *EdD Dissertation* Marymount University, Arlington County, VA

[30] Rodriguez R D 2022 Components of Latinx parental involvement in high school *EdD Dissertation* Texas A&M University—Kingsville

[31] Ivie R, Czujko R and Stowe K 2002 Women physicists speak *AIP Report* www.aip.org/statistics/reports/women-physicists-speak (Accessed: 28 June 2024)
Ivie R and Guo S 2006 Women physicists speak again *AIP Report* https://ww2.aip.org/statistics/women-physicists-speak-again (Accessed: 7 June 2024)

[32] Archer L, Moote J, Francis B, DeWitt J and Yeomans L 2017 The 'exceptional' physics girl: a sociological analysis of multimethod data from young women aged 10–16 to explore gendered patterns of post-16 participation *Am. Ed. Res. J.* **54** 88–126

Chapter 6

Education

She was the one who told me that I could make it even though I was a woman, and she did warn me that the road ahead for women in science might be more difficult, but not to be deterred.
—Millie Dresselhaus, on advice from Nobel Laureate Rosalind Yalow [1]

We have seen (chapter 4) that the proportion of women in physics in the English-speaking world is much lower than 50% at nearly every stage of education and career. What factors are causing this? Research on women and science has been conducted for over fifty years, and despite all we have learned, we are still discovering important new factors that play into this under-representation.

The data presented in the previous chapter (chapter 5) support the idea that the split between girls' and boys' participation levels in physics happens very early, prior to college. The research literature supports this conclusion. This chapter will explore some of the factors in early life and education that start to differentiate the genders in their attitude towards physics. Readers are encouraged to remember that gender differences which are noted are simply observations, and not intended as a statement of value. Whether a difference preferences or advantages one group over another is something readers should consider for themselves, along with the underlying biases and beliefs that lead us to these conclusions.

6.1 Early science education

One of the problems facing US science education and female participation is the lack of good science in elementary schools. Many elementary teachers are afraid of science [2] and will avoid it where possible. The issue is serious enough that there is a Science Teaching Anxiety Scale [3]! Only about one third of kindergarten–grade 5 teachers have any coursework in physics, and only 17% felt well prepared to teach physical science [4]. In one study, preservice elementary teachers in the US had the least self-efficacy about teaching physics, among the sciences [5]. In the UK, half of the primary teachers in a survey noted lack of confidence and ability to teach science [6]. When they do teach it, their attitudes are communicated to the children they are teaching, with sometimes negative consequences [7]. Standards at the state and national level can help somewhat [8], but if the teachers are uncomfortable with, afraid of, or poorly educated in science, they cannot serve the best interests of their students, or of our society in general. One study [9] notes that female elementary science teachers' fear of math is transferred to the girls in their class, though not to the boys. Since most primary school teachers are women, this presents a major problem.

A related issue is that when teachers are uncomfortable with science, they tend to fall back on lecturing and rote learning, which is what they likely experienced in their college science courses [10]. They also may or may not have experienced a science methods course that focused on active learning. If primary teachers use an active, engaged teaching style for most topics, but a more rote and passive style for science, they will likely be creating negative attitudes towards science in their students.

A fascinating task that has been used for decades to determine ideas about scientists is the Draw A Scientist Test [11]. Participants are asked to take a few minutes to draw a scientist—usually no other information is given. Milford and Tippett [12] found that preservice elementary teachers were likely to draw highly stereotypical scientists: male, lab coat, glasses, crazy hair. These stereotypes of what a scientist looks like are likely to be transferred to their students. This test has been very consistent over the last few decades in drawing out stereotypes of scientists, from elementary students to university students [13].

One of the major problems of elementary school science teachers who are afraid of science comes from the fact that they serve as such an important role model for both the girls and boys in their classrooms at an important point in social and cultural development. But this problem also represents an opportunity. By combating negative attitudes towards science among these teachers, we can not only help in the transfer of more positive attitudes, but also create better role models for females who like science.

6.2 Extracurricular and informal science

Science camps are prevalent across most Anglophone countries, and science camps for girls are growing in number. Why the push for such opportunities for girls? Because having experiences with science as a child can change attitudes towards science and motivation to choose science as a career later in life. While the research on these programs is mixed [14], most are providing important pieces [15] in the larger puzzle of attracting girls to science. These camps often bring such factors to bear on the issue as mentors or counselors who work with the girls, a sensitivity to the cultural background of the girls being recruited, and peer social interactions. These camps target girls ranging from young elementary school to high school (see Valla and Williams [16] for a thorough listing of articles on this topic). Another advantage of non-school science activities such as camps is that girls can start to see science in their world, and their world as full of science [17]. Camps can be useful for improving interest, confidence, and self-efficacy [18]. Though an important factor to remember when constructing programs for this sort of program is if informal science experiences are built on stereotypical lines, whether intentionally or not, they can exacerbate the behaviors of young children along these lines [19].

Extracurricular activities show a correlation with likelihood to pursue a STEM career for groups of year 11 and year 12 pupils, while girls were much less likely to participate in these activities [20]. So let's get them participating!

Besides participating in such activities, having young women facilitate or lead such activities can provide specific benefits: improved confidence and performance beliefs as well as help for development of physics identity [21].

6.3 High school physics education

High school physics is one of the first places where students have the option to choose or drop out of physics. Because physics is commonly taught in the US during the last year of high school, it is often elective (as with A-level physics), with many students allowed to skip senior year science altogether, or choose another science instead of physics. This is a major inflection point for losing women in physics. For those that do choose to take a physics course in high school, what experiences are young women having in those classes? And what impacts those experiences?

In the US, about two-fifths of high school students will have a female physics teacher [22]. This number is rising, but very slowly. In New Zealand the ratio is similar, with about 200 female physics teachers and 500 male physics teachers, as is Ireland's ratio at 40% [24]. Wales and Scotland have between one quarter and one third female physics teachers [25]. In Australia only about a quarter of physics teachers are women [26]. Young women have no guarantee of seeing a female role model in their secondary physics courses.

The gender of one's high school physics teacher can have interesting impacts, although the data on exactly how those effects play out are open to interpretation. A possible silver lining is that whether or not a high school physics teacher is of the same gender as the student is not a factor that is likely to produce stronger science identity in college [27].

Students in high school physics classes are more likely to rate the performance of a female physics teacher lower than a male physics teacher. This was true for both the young men and young women in the class studied [28]. This is consistent with studies of student evaluations of teachers at the university level that show that women faculty are consistently rated lower than male faculty [29], even in one of the countries with the highest gender equality ratings in the world [30]. Using different criteria, Gilmartin [31] found that the percent of female faculty in high school science classrooms in a small study did not correlate to how the students talked about their teachers, nor did students refer to teachers as role models.

Grades in secondary physics show mixed results. GCSE scores in parts of the UK show boys outperforming girls at the top scoring levels: 14.2% of boys and 10.7% of girls earned a 9, the highest grade, while 45.7% and 41.8% earned a 7, 8 or 9 [32]. The elective A-level in physics shows a very small gap, with 8.8% of boys and 8.5%

of girls earning the highest grade. For US students, female students have higher grades generally, including in physical sciences [33]. Though they do more poorly on the science section of standardized tests [34].

Too few teachers encourage students to study physics. A Canadian research study [35] of undergraduate students with research awards shows that only 32% of the women in this population had a high school teacher who encouraged them to continue in science and engineering, and the males were only slightly higher at 37%. And these were students with research awards in science and engineering! Imagine what it is like for less obviously science-inclined students. And it's consistent across countries. Mujtaba and Reiss [36] note that 'girls are less likely to be encouraged to continue with physics' after it is no longer compulsory in the UK.

This may be partly due to teacher bias regarding girls and science. Because of our societal biases against females in science, both teachers and students can have biases that boys do better in science and the sciences are more masculine fields [37]. Encouragement, or lack thereof, isn't just a teacher phenomenon: as noted in chapter 5, this extends to parents. And girls in secondary physics courses not only receive less teacher encouragement than their male peers, but they also receive less *peer* encouragement [38].

While we are in the cultural process of eliminating much of the most explicit discouragement of girls in the educational setting, we still regularly hear about blatant sexism in high school classes and counseling rooms [39]. Without encouragement from academic advisors and teachers, it is difficult for young women to push through the barriers and move into science. '[I]t is crucial for career counselors to develop interventions that build minority and female students' confidence and increase their self-understanding [40].'

The physics curriculum itself can also be biased in regards to mentions of scientists within concepts (Ohm's law, Snell's law, Newton's laws) [41]. A study of the high school science curriculum across Australian states found that physics had the highest mentions of scientists both alone and within concepts. Only one female scientist was mentioned in any course: Rosalind Franklin. The authors also note a strong Eurocentric bias. Similar results were found in Ireland [42]. Since many of our physical laws were named for scientists pre-1900, it is not a big surprise, and a historical focus in a course will exacerbate this issue [43]. Perhaps we can start renaming laws for their actual content? Kirchoff's loop law could become the circuit loop law. Something to consider.

The biggest boom in the literature in the last decade is surely centered on science identity and sense of belonging. Chapter 3 gives an overview of these concepts. There are many more articles than I could hope to list here. How does science identity play out in the physics classroom? (This will include secondary and tertiary education data.)

Science identity comprises three parts: performance, recognition, and competence [44]. In physics, interest can play a fourth role [45]. Recognition probably plays the largest part [46]. Others speaking of you as a physics person is obviously going to be a part of identifying yourself as a physics person. In terms of recognition, seeing others like you in a field makes it easier to believe you are part of that in-group [47].

Consuming STEM media can help promote STEM identity, specifically TV and videogaming [48].

When does science identity start to differ for boys and girls? Typically in high school [49]. Strong science identity is a good predictor of staying in science, for men and women [50]. In a gender-imbalanced field such as physics, the simple fact of being one of the few (or the only) girls in a class can inhibit making connections between oneself and the field [51]. And sexist behavior of men in a male-dominated class can damage women's social identities [52]. Unfortunately, having more women in a physics class doesn't necessarily make for a higher adoption of physics identity [53].

Sense of belonging is a related concept, and also predicts intentions to stay in a field [54]. Not only that, but it can predict academic performance in physics [55]. It really is what it says: do you feel like you belong in this field/place/group? For female physics students at university, a sense of belonging had a strong connection to physics identity for seniors but not first-year undergraduates [56]. As with science identity, there is so much more out there to read, and I strongly encourage those who are interested to dig into these two areas.

Does the way a classroom is run make a difference? Yes. Math preparation in high school is important for both young men and women for college science [57]. We know that high school physics pedagogy and activities can affect students' decision to continue in physics, and for young women in particular [58]. Specifically, courses that focus on understanding over memorization helped women more than men. And doing problems requiring long written responses was negatively associated with physics persistence for women.

For decades, single-sex classrooms [59] have been proposed as one solution to the gender divide in science. The literature on this subject has been quite variable [60] over the years, and remains so. Studies suggest 'that girls in single-sex classes have a better self-concept of ability in masculine school subjects than the girls in mixed-sex classes because gender-related self-knowledge is less accessible once the opposite sex is absent' [61], but also '[w]hile girls-only schools appear to foster more participation in physical science courses or to encourage more interest in physical careers among their students, these differences are attributable to factors other than gender compositions of schools.' [62] The IOP in the UK did find that single-sex classrooms were better at countering gender differences in progression to A-level physics [63], while another study found single-sex instruction can help high-ability girls [64]. An Australian study noted that females in single-sex schools were significantly more likely to specialize in physics, compared to their mixed school peers [65].

Same-sex instruction may not have huge effects, though; a 2023 study on discourse in single-sex high school classes found that boys' and girls' discourse did not differ significantly [66], while an Australian look at girls in same-sex or coed physics classes found very few differences in motivation [67]. A more recent study found that girls from single-sex schools did not choose physical or life sciences in college at higher rates than those from coed schools [68]. I suspect that the data on single-sex schooling are in significant flux.

One common critique of single-sex classes is that they can serve to falsely insulate women from the harshness of the broader science community, thus putting them at a disadvantage later when they enter the world beyond the classroom. As Gertrude Elion noted:

> I hadn't been aware that there were doors closed to me until I started knocking on them. I went to an all-girls school. There were 75 chemistry majors in that class, but most were going to teach it …. When I got out and they didn't want women in the laboratory, it was a shock …. It was the Depression and nobody was getting jobs. But I had taken that to mean nobody was getting jobs … [when I heard] 'You're qualified. But we've never had a woman in the laboratory before, and we think you'd be a distracting influence.' [69]

Perhaps the best comment on single-sex classrooms comes from Michael Neuschatz, at the American Institute of Physics in 1995: 'single sex classrooms cannot substitute for confronting head-on gender prejudice in physics education' [70].

Intervention programs can support an improved attitude towards physics and a reduction in stereotypical views of physics and gender. The IOP in Britain has completed several 'Improving gender balance' projects with the goal to improve the number of girls that choose physics in secondary school [71]. The projects have been successful in raising awareness of gender bias among teachers. Providing counter-narratives in high school classrooms can also help reduce bias in career choice among young women [72].

One potential educational intervention on the gender gap in many areas is found in the research on intellectual mindset [73]. Some people believe that intelligence is a fixed trait—that a person is smart or not. Others believe that intelligence is malleable, and one can change and develop how smart you are. Fixed-mindset students tend to stop trying when they fail at a task, saying 'I am not good at physics'. Growth-mindset students will continue to try, believing that they can *become* good at physics [74]. What makes this so important as a tool is evidence that you can teach students to shift to a growth mindset [75]. This is hopeful, as the main researcher for mindset has stated that a growth mindset can help increase girls' confidence in STEM areas [76]. We should temper our hopes, however, since a meta-analysis in 2023 found that most effects were very small or statistically non-significant [77].

Readers should also keep in mind that the majority of this research was conducted on and with white girls, and that has the potential to create impressions that may not be generalizable to all female populations. 'Developmental models that may be appropriate in certain white, middle- and upper-middle-class settings cannot be generalized to all girls in all contexts.' [78] We also need to remember that every human is individual, and the differences among women are usually bigger than the differences between genders [79].

6.4 Undergraduate physics education

At the undergraduate level, physics is an elective course, unlike math or English for most students. Aside from a very small number taking the course for general education/distribution credits, only those with an interest in science or a career choice related to science will take physics classes. In conjunction with all of the other factors that lead women to opt out of a physics path, this creates an environment in which women are only about 25% of college physics students—so one of the first things a young woman might notice when she takes physics is the relative lack of fellow women in her classes.

The next thing a young woman in physics might notice is how few female faculty members there are. Only 14% of physics faculty in the US are women, and with only about 20% of graduate students being women, if they have a Teaching Assistant, there is only a 1 in 5 chance of that TA being female [80]. As with the high school classes, this lack of role models is a subtle, often subconscious, message that physics may not be a welcoming place for women. In the UK the numbers are lower still, with only 9% at the professor level and 19% at lecturer level [81]. A natural experiment at the US Air Force Academy exists, with students randomly assigned to sections of classes. For introductory STEM courses, women with a female instructor were more likely than women with a male instructor to choose a STEM major [82].

Women's self-confidence [83] and self-efficacy [84] are often lower or different than men's in the physics classroom. This confidence can affect persistence in pursuing physics and career choices [85]. As mentioned above, sense of belonging and science identity also affect women's likelihood of continuing in physics. Interestingly, one study found that physics identity was lowest in the fourth year of undergraduate studies for both men and women [86]. Recognition from peers is one important part of building a physics identity; yet studies show that men are more likely to be recognized as high performers [87], even when women get better grades [88]. Peer mentors as well as faculty mentors can help with a woman's self-confidence in physics, promoting a sense of belonging [89].

In her book *Delusions of Gender*, Fine spends most of a chapter on women in male-dominated fields [90]. She notes one study looking at how undergraduate women responded to a study showing higher math scores for males on the SAT; the women were less likely to identify with certain female characteristics. 'Parts of their identity were being hurled overboard in an attempt to remain afloat in male-dominated waters.' [90]

Data from Australia are limited for physics, but one study of science undergraduates found the same results as US and UK researchers: high levels of science identity and belonging mean better retention for women [91]. A recent systematic review of Australian gender differences in undergraduate STEM noted that there is a lack of research focusing on Australia [92]. In their study of 36 papers, women's lower self-efficacy was the most popular topic, along with learning styles, motivation, culture, and science identity.

Laboratory work can be a blessing or a curse for young women in a physics class. While practical work can help promote self-efficacy for women [93], it often is a place where women are relegated to notetaking instead of working with equipment [94]. One study on in-person versus online physics labs found that controlling for type of lab, there were no significant gender differences in help-seeking beliefs [95]. In group and lab work, while women typically out-perform men, they are viewed as less able, in a study of life and physical sciences [88].

Some studies have found that women prefer theoretical physics classwork to practical/lab work [96]. This may be connected with the behavior of girls in labs, but which way causation goes we don't know.

A fascinating look at men and women's views of roles and group composition in undergraduate physics labs found surprisingly few gender differences [97]. Women much preferred working together on tasks (as opposed to taking turns or individuals doing different tasks), as did men, though women's preference for collaborative work was about 30% higher. Men had a preference for analyzing data, while women had a preference for notetaking and group management. Most of the men and women had no preference for group gender composition (78% and 73% respectively), though women were more likely to prefer majority female-identifying groups. For a physics class with non-majors, women were more likely to prefer notetaking, but the preference for sharing tasks was not found [98].

A similar preference for sharing tasks was found by Doucette, along with findings that suggest avoiding having one isolated female in lab groups [99].

As with high school, the pedagogy and curriculum of the college classroom can have an effect on women's intent to continue with physics. There is so much information here that I could never begin to address it all. A sampler: There are gender gaps on the major conceptual tests [100], procrastination [101], grades in different parts of a course [102], scaffolding in problem solving [103], online homework [104], and active-learning or traditional/lecture type of course [105]. If this area interests you, start by heading over to *Physical Review Physics Education Research*, an open-access journal.

There are so many issues affecting how women interact with the physics classroom that it is hard to identify the individual factors which are correlated with the biggest differences in performance; instead, we see a 'smog of bias' [106]. Nor are solutions obvious.

One of the issues that has been problematic for girls and women in physics classes that I am particularly interested in is the historical practice of using contexts that are more stereotypically or historically male than female. Most of the data are older, but I expect still relevant. Projectile motion problems might include baseball, while a frictionless surface might be ice on a hockey rink. Centering ballistics and military contexts, torque wrenches [107], and footballs can serve as subtle messages that physics is a man's field.

Rennie and Parker studied the effects of context on gender and performance for several years [108]. They found that the context of a question can affect how boys and girls perform on a problem. In physics the effect of context on the problems of the Force Concept Inventory [109] has been studied, and it has been shown that men and women perform differently with different contexts [110].

A British head of physics at a girls' school notes: Girls love practical work. They engage with everyday examples, but it depends on the context because topics such as cars might switch them off. The teacher should let them construct the relationship with the topic, themselves, from practical work. [111]

Being conscious of the contexts we use can help promote a more welcoming environment for the young women and girls that are engaging with physics.

Similarly, certain test formats may favor males. There is less current research, but concrete problems, visualizations, and specific content may be part of a male bias on certain tests [112]. But the data are mixed; see Gladys *et al* [112] for a good review.

Another point to remember is that many of the tests and assessments used in our educational system were only pilot tested on classrooms of boys or men. From the SAT to the GRE to many commonly used tests of physics, if women perform more poorly than men, we cannot know without careful study if that is a fault of the women's understanding of the material, or if the design of the system and of tests and classes is *actually* biased towards men. As mentioned in the introduction, we need to be careful not to set men's performance as the standard and measure women against an inappropriate measuring stick. Jane Austen notes in her book *Persuasion*: 'I do not think I ever opened a book in my life which had not something to say upon woman's inconstancy. Songs and proverbs all talk of women's fickleness. But...these were all written by men.' If women underperform on an exam that was written by men, tested on men, and optimized for men, we should not immediately blame the women.

College students, like high school students, hold a bias against female physics teachers, as demonstrated in an experiment [113] where four professors delivered identical videotaped lectures. Students watched one of the videos, then rated the teachers. Male students showed a significant bias in favor of male teachers. Female students were more equitable in their evaluations except for rating female teachers higher in communication skills and *lower* in scientific skills. In an Australian study of half a million evaluations, female teachers in science have a significant chance of lower evaluations than male teachers [114]. The bias may be worse for those more vested in the field: a strong physics identity among high school students was related to a larger bias towards male teachers [115].

A fascinating US study connected the idea of brilliance and student evaluations of professors using the website RateMyProfessor [116]. Fields where students used the words 'brilliant' or 'genius' more frequently were fields with lower numbers of women and African Americans.

A mindset study for an introductory calculus-based physics course found that by the end of the course, all students had *more* gendered views of physics in the realm of needing to be smart to do physics [117]. They also found that only one dimension of mindset (My Ability) correlated with course grade.

One thing that has improved over the last few decades is diversity of representation of people in physics textbooks [118]. As recently as twenty years ago, if there were people pictured, those people were likely to be (a) male and (b) doing something stereotypically male. Today's texts are much more diverse in their pictures, both in terms of people and activities. Larsen [119] studied four decades of gender inclusive pictures in astronomy texts and found a significant increase in the number of women included over time. Physics texts still tend to show white males, with limited improvement in the last twenty years [120]. For interested readers, UNESCO [121] has produced a guide to gender equality and textbooks.

Although how much do textbooks have to do with promoting interest in the sciences? We don't know. 'I worry that we increasingly teach science out of textbooks. Someone once said that's like learning to drive a car from reading of a book. It's the practical science that really turned me on, doing it myself.' Dame Nancy Rothwell [122].

As one reaches tertiary education, the role of the advisor grows in importance. And just as with secondary school advisors, college advisors often (knowingly or unknowingly) discourage women from physics and other STEM majors [123]. Though we no longer expect to hear 'Why do they expect me to teach calculus to girls?' [124], we still have to work against unconscious biases where young women are told to take biology instead of physics, or to stop taking math classes. More obvious sexism, while rarer, still exists, such as the comment from Nobel Prize awardee Tim Hunt in June 2015, that he has problems when 'girls' are in the lab: 'You fall in love with them, they fall in love with you and when you criticize them, they cry.' [125] While he did apologize somewhat for this statement, that anyone feels this is an appropriate thing to say in 2015 shows how much work we have to do.

Along with outright sexism, sexual harassment is still a factor for women in physics classrooms and laboratories. Recent news articles [126] about sexual harassment in the sciences demonstrate how common this problem still is. A study in the US found a most alarming rate of 74% of female physics majors experiencing some sort of harassment [127]. Most universities and many large departments now have policies on sexual harassment, and there are some universities with online training programs to help prevent sexual harassment. Some schools have new students take training on sexual consent [128]. Despite these advances, it's clear that we still need to be aware of this issue, and all schools need to have programs in place to support victims of sexual harassment and sexual abuse. Firm statements from agencies such as the US National Science Foundation [129] can also help move the culture forward.

Getting involved in research as an undergraduate can be an extremely rewarding experience [130], and it makes a student more attractive to graduate schools. Because of the high value of undergraduate research, there are conferences devoted to supporting undergraduate women's research. The Conferences for Undergraduate Women and Gender Minorities in Physics are run by the American Physical Society [131]. Their purpose is

> to help undergraduate women continue in physics by providing them
> with the opportunity to experience a professional conference, informa-
> tion about graduate school and professions in physics, and access to

other women in physics of all ages with whom they can share experiences, advice, and ideas. [131]

A similar conference has been developed for the UK and Ireland [132].

Research can help us move forward in our goal. Along with countering bias and stereotypes, we can look at how women enter science rather than just what causes them to leave. A study that focused on the life course of scientists encourages policies and efforts to 'facilitate the flow of high school students who expected non-S/E [science/engineering] college majors into S/E majors during the first year of college' [133].

6.5 Postgraduate physics education

The experiences that women have in postgraduate physics tend to be harsher and more hostile than during their undergraduate years [134]. Despite this, there is only a small drop in participation from master's degrees to doctoral degrees, possibly because the women who have made it this far are solidly committed to the field or have evolved effective coping mechanisms for dealing with bias. While some women who go in for a PhD do drop out with a master's degree, this happens at a similar rate to men leaving the field, at least in the US.

The issues facing female graduate students are broadly similar to those of undergraduates, but with some additional areas where bias can enter into the process. Because graduate school is focused so strongly on research, the relationship a student has with their advisor is extremely important [135]: 'Without an advisor who is willing to encourage and direct, women are often unable to puzzle out the strategies necessary to get through graduate school'. Some advisors are not good at advising and mentoring women, while others simply don't want to work with women at all [136].

Here's what one female graduate student had to say about her first advisor, who never let her finish a sentence. She felt that his message was 'I'm going to regard you, graduate student, as nothing' [136]. Such messages are not exclusive to male advisors. Female advisors can be just as poor at encouraging women as men. Informal networks can be very helpful for female graduate students in these situations, allowing information to be passed along to new graduate students about who is a good or bad advisor. I benefited from this in my own career, as second- and third-year female students told the first-year female students which faculty members to avoid when I was in graduate school.

Advisors also determine how much a graduate student does in the lab. A poor advisor may give too much work, provide too little support, or use very poor communication. Students who are poorly served in this way will not acquire many of the skills that the PhD system is intended to develop. On the other hand, some advisors may be over-protective of their students [137]. These advisors are likely to spoon-feed their students, or fail to challenge them, and to limit the opportunities they allow their students. The strongest PhD candidates on the job market are those with a breadth of experience.

Advisors can be part of the reason women leave graduate school before earning their degree [138]. They can also be an important part of developing social networks in the field [139]. Picking a research group that a woman feels comfortable in may mean choosing a group that is not her first (or second) choice [140]: 'My biggest concern going forward is that I want to feel okay to take up space in lab.'

Good advisors will encourage and challenge their students, while providing enough support to enable students to succeed. Having graduate students assist in grant writing, serve as first author on papers, give conference presentations, and participate in professional development such as workshops and summer schools are all things that will help not just women but all students. A supportive advisor can be male or female [141]: 'Women and men faculty do not, simply by virtue of their gender, automatically make good or poor mentors for female students [142].'

Having a supportive advisor or mentor can have long-reaching consequences. Louis Leakey (who worked with his wife Mary) believed that women would be better at studying primates in their natural habitat, and from this were born the leaders of three fields of primatology, the 'trimates': Jane Goodall, Dian Fossey, and Birute Galdikas [143].

Graduate students are also more sensitive to the climate and culture of a department, because unlike an undergraduate, their whole education is now confined to one department, often to one building. The term 'chilly climate' has been around since the 1980s [144], and it is particularly appropriate to apply in a graduate school environment.

As with undergraduates, women in physics graduate school are susceptible to lowered self-efficacy and feelings of belonging. This confidence can affect persistence in pursuing physics [145] and career choice. Being a graduate student doesn't necessarily mean you have a strong science identity, since being a researcher may be a stronger identity [146].

Luckily, women entering physics graduate school today have several resources to help determine the climate of a department they are considering. Project JUNO, part of the IOP, had a goal 'to recognise and reward departments that can demonstrate they have taken action to address the under-representation of women in university physics and to encourage better practice for both women and men' [147]. While JUNO closed in 2023, the IOP is working on another initiative. In the US, the APS has a site visit program [148] to help departments find out how to better the climate for their female and racial minority colleagues. And there is the growing SEA Change initiative out of the American Association for the Advancement of Science [149].

A related issue for graduate schools has to do with the Physics GRE, the standardized test in the US for graduate school. It came under attack in the 1990s as detrimental to the field's gender diversity [150]. One study of the physics GRE found it did not correlate with PhD completion, and it limited access to graduate schools by under-represented groups [151]. Looking at undergraduate GPA or the physics GRE, only the grades predict PhD completion [152].

In terms of climate, women in postgraduate education have to be careful of how they dress, how they speak, where they hang out [44]. Graduate school is about becoming a physicist, in many ways. 'Break down the ego ... until it is a barely visible puddle on the floor, then build it up again in the image of the professor.' [153] Do physicists wear dresses? 'I know a girl who dresses in skirts and heels when at summer schools, but told me she never dresses up around her supervisor, so that he will think she is always working.' [154] For some reason there is an idea that wearing nice dresses means you are a bad scientist. As one graduate student shares: 'At a conference, [my advisor] pointed out another student presenter to us who was wearing a pretty dress. He said, 'look at what that chick's wearing, she obviously doesn't know what she's talking about!'' [155].

In my own experience as a graduate student, I was walking up the stairs in the physics building, holding my long skirt in my hand so as not to trip. A male faculty member stopped in the middle of the stairs, looked puzzled, and told me that he didn't think he had ever seen such behavior in that building before. The message this sent was quite clear, even if the professor in question did not intend it explicitly: skirt $\neq$ physicist.

From the home hearth to the classroom seat to the lab bench, women experience physics differently than men do. These differentiated experiences lead to different attitudes and beliefs about their relationship to physics. For some women, this difference is enough to make them leave physics altogether. Other women stick it out, and move on to PhDs and jobs. They are the successes of the educational system. In the next chapter we will explore how working women encounter the culture of physics.

References

[1] US Department of Energy 2012 Fermi award winners: Q&A Dr Mildred S Dresselhaus *Office of Science* https://science.osti.gov/Science-Features/News-Archive/Featured-Articles/2012/06-06-12 (Accessed: 24 June 2024)

[2] Riegle-Crumb C, Morton K, Moore C, Chimonidou A, Labrake C and Kopp S 2015 Do inquiring minds have positive attitudes? The science education of preservice elementary teachers *Sci. Educ.* **99** 819–36
Tosun T 2000 The beliefs of preservice elementary teachers toward science and science teaching *School Sci. Math.* **100** 374–79
Kazempour M 2014 I can't teach science! A case study of an elementary pre-service teacher's intersection of science experiences, beliefs, attitude, and self-efficacy *Int. J. Environ. Sci. Educ.* **9** 77–96

[3] Novak E, Soyturk I and Navy S 2022 Development of the science teaching anxiety scale for preservice elementary teachers: a Rasch analysis *Sci. Educ.* **106** 739–64

[4] Trygstad P 2013 *2012 National Survey of Science and Mathematics Education: Status of Elementary School Science* (Chapel Hill, NC: Horizon Research)

[5] Al Sultan A 2020 Investigating preservice elementary teachers' subject-specific self-efficacy in teaching science *EURASIA J. Math. Sci. Tech. Educ.* **16** em1843

[6] Murphy C 2008 Primary science teacher confidence revisited: ten years on *Educ. Res.* **49** 415–30

[7] Jarrett O 1999 Science interest and confidence among preservice elementary teachers *J. Elem. Sci. Educ.* **11** 47–57

Beilock S, Gunderson E, Ramirez G and Levine S 2010 Female teachers' math anxiety affects girls' math achievement *Proc. Natl Acad. Sci.* **107** 1860–3

[8] The three dimensions of science learning *Next Generation Science Standards* http://nextgenscience.org/

[9] Beilock S, Gunderson E, Ramirez G and Levine S 2010 Female teachers' math anxiety affects girls' math achievement *Proc. Natl Acad. Sci.* **107** 1860–3

[10] Bergman D and Morphew J 2015 Effects of a science content course on elementary preservice teachers' self-efficacy of teaching science *J. Coll. Sci. Teach.* **44** 73–81
Varma T, Volkmann M and Hanuscin D 2009 Preservice elementary teachers' perceptions of their understanding of inquiry and inquiry-based science pedagogy: influence of an elementary science education methods course and a science field experience *J. Elem. Sci. Educ.* **21** 1–22

[11] Chambers D 1983 Stereotypic images of the scientist: the draw a scientist test *Sci. Educ.* **67** 255–65

[12] Milford T and Tippett C 2013 Preservice teachers' images of scientists: do prior science experiences make a difference? *J. Sci. Teach. Educ.* **24** 745–62

[13] Thomas M, Henley T and Snell C 2006 The draw a scientist test: a different population and a somewhat different story *Coll. Stud. J.* **40** 140–8

[14] Hughes R, Nzekwe B and Molyneaux K 2013 The single sex debate for girls in science: a comparison between two informal science programs on middle school students' STEM identity formation *Res. Sci. Educ.* **43** 1979–2007
Bhattacharyya S, Nathaniel R and Mead T 2011 The influence of science summer camp on African-American high school students' career choices *School Sci. Math.* **111** 345–53

[15] Levine M, Serio N, Radaram B, Chaudhuri S and Talbert W 2015 Addressing the STEM gender gap by designing and implementing an educational outreach chemistry camp for middle school girls *J. Chem. Educ.* **92** 1639–44
Gándara P and Bial D 2001 Paving the way to postsecondary education: K-12 intervention programs for underrepresented youth *Report* National Postsecondary Education Cooperative Access Working Group, US Department of Education, National Center for Education Statistics http://nces.ed.gov/pubs2001/2001205.pdf (Accessed: 29 June 2024)
Valla J and Williams W 2012 Increasing achievement and higher-education representation of under-represented groups in science, technology, engineering, and mathematics fields: a review of current K-12 intervention programs *J. Women Minor Sci. Eng.* **18** 21–53

[16] Valla J and Williams W 2012 Increasing achievement and higher-education representation of under-represented groups in science, technology, engineering, and mathematics fields: a review of current K-12 intervention programs *J. Women Minor Sci. Eng.* **18** 21–53

[17] Gonsalves A, Rahm J and Carvalho A 2013 We could think of things that could be science': girls' re-figuring of science in an out-of-school-time club *J. Res. Sci. Teach.* **50** 1068–97

[18] Farland-Smith D 2016 My daughter the scientist? Mothers' perceptions of the shift in their daughter's personal science identities *J. Educ. Issues* **2** 1–21

[19] Hazari Z, Dou R, Sonnert G and Sadler P 2022 Examining the relationship between informal science experiences and physics identity: unrealized possibilities *Phys. Rev. Phys. Educ. Res.* **18** 010107

[20] Siani A and Dacin C 2018 An evaluation of gender bias and pupils' attitude towards STEM disciplines in the transition between compulsory and voluntary schooling *New Dir. Teach. Phys. Sci.* **13**

[21] Randolph J, Perry J, Donaldson J P, Rethman C and Erukhimova T 2022 Female physics students gain from facilitating informal physics programs *Phys. Rev. Phys. Educ. Res.* **18** 020123

[22] White S and Chu R Y 2024 Who teaches high school physics? *AIP Report* https://ww2.aip.org/statistics/who-teaches-high-school-physics-19 (Accessed: 3 June 2024)

[23] Association for Women in the Sciences 2011 Women in science: a 2011 snapshot *Booklet* www.awis.org.nz/assets/Files/AWIS-Stats-2011-Booklet.pdf (Accessed: 29 June 2024)

[24] Chormaic S, McLoughlin E and Gunning F 2005 The current situation of women in physics in Ireland *AIP Conf. Proc.* **795** 133–4

[25] The Royal Society 2007 The UK's science and mathematics teaching workforce *Report* https://royalsociety.org/topics-policy/projects/state-of-nation/teaching-workforce/ (Accessed: 29 June 2024)

[26] Weldon P R 2015 The teacher workforce in Australia: supply, demand and data issues *Policy Insights* issue 2 (Melbourne: ACER)

[27] Chen C, Sonnert G and Sadler P 2019 The effect of first high school science teacher's gender and gender matching on students' science identity in college *Sci. Educ.* **104** 75–99

[28] Potvin G, Hazari Z, Tai R and Sadler P 2009 Unraveling bias from student evaluations of their high school science teachers *Sci. Educ.* **93** 827–45

[29] Boring A, Ottoboni K and Stark P 2016 Student evaluations of teaching (mostly) do not measure teaching effectiveness *Sci. Open Res.* **0** 1–11

[30] Sigurdardottir M S, Rafnsdottir G L, Jonsdottir A H and Kristofersson D M 2022 Student evaluation of teaching: gender bias in a country at the forefront of gender equality *Higher Educ. Res. Dev.* **42** 954–67

[31] Gilmartin S, Denson N, Li E, Bryant A and Aschbacher P 2007 Gender ratios in high school science departments: the effect of percent female faculty on multiple dimensions of students' science identities *J. Res. Sci. Teach.* **44** 980–1009

[32] Durk J, Davies A, Hughes R and Jardine-Wright L 2020 Impact of an active learning physics workshop on secondary school students' self-efficacy and ability *Phys. Rev. Phys. Educ. Res.* **16** 020126

[33] Westrick P 2018 Examining precollege gender differences and similarities among fourth-year undergraduate students, with a focus on STEM *Technical Brief* ACT Research and Policy https://act.org/content/dam/act/unsecured/documents/R1668-precollege-gender-differences-2018-01.pdf (Accessed: 13 June 2024)

[34] Buddin R 2014 Gender gaps in high school GPA and ACT scores *Information Brief* ACT Research and Policy https://act.org/content/dam/act/unsecured/documents/Info-Brief-2014-12.pdf (Accessed: 13 June 2024)

[35] National Sciences and Engineering Research Council of Canada 2017 Women in science and engineering in Canada *Report* https://nserc-crsng.gc.ca/_doc/Reports-Rapports/WISE2017_e.pdf (Accessed: 30 June 2024)

[36] Mujtaba T and Reiss M 2014 A survey of psychological, motivational, family and perceptions of physics education factors that explain 15-year-old students' aspirations to study physics in post-compulsory English schools *Int. J. Sci. Math Educ.* **12** 371–93
Mujtaba T and Reiss M 2013 Inequality in experiences of physics education: secondary school girls' and boys' perceptions of their physics education and intentions to continue with physics after the age of 16 *Int. J. Sci. Educ.* **35** 1824–45

[37] Hand S, Rice L and Greenlee E 2017 Exploring teachers' and students' gender role bias and students' confidence in STEM fields *Soc. Psych. Educ.* **20** 929–45

[38] Mujtaba T and Reiss M 2013 Inequality in experiences of physics education: secondary school girls' and boys' perceptions of their physics education and intentions to continue with physics after the age of 16 *Intl. J. Sci. Educ.* **35** 1824–45

[39] Sadker M, Sadker D and Zittleman K 2009 *Still Failing at Fairness* (New York: Scribner)

[40] Mau W-C 2003 Factors that influence persistence in science and engineering career aspirations *Career Dev.* Quart. **51** 241

[41] Ross K, Galaudage S, Clark T, Lowson N, Battisti A, Adam H, Ross A and Sweaney N 2023 Invisible women: gender representation in high school science courses across Australia *Aust. J. Educ.* **67** 231–52

[42] Pillion K and Bergin S 2022 The representation of women in Irish Leaving Certificate Physics textbooks *Phys. Educ.* **57** 025017

[43] Keast V 2022 Gender bias in New South Wales Higher School Certificate (HSC) physics *Aust. J. Educ.* **66** 26–39

[44] Carlone H B and Johnson A 2007 Understanding the science experiences of successful women of color: science identity as an analytic lens *J. Res. Sci. Teach.* **44** 1187–218

[45] Hazari Z, Sonnert G, Sadler P and Shanahan M C 2010 Connecting high school physics experiences, outcome expectations, physics identity, and physics career choice: a gender study *J. Res. Sci. Teach.* **47** 978–1003

[46] Whitcomb K, Maries A and Singh C 2023 Progression in self-efficacy, interest, identity, sense of belonging, perceived recognition and effectiveness of peer interaction of physics majors and comparison with non-majors and PhD students *Res. Sci. Educ.* **52** 525–39
Godwin A, Potvin G, Hazari Z and Lock R 2015 Identity, critical agency, and engineering majors: an affective model for predicting engineering as a career choice *J. Eng. Educ.* **105** 312–40

[47] Merritt S, Hitti A, Van Camp A, Shaffer E, Sanchez M and O'Brien L 2021 Maximizing the impact of exposure to scientific role models: testing an intervention to increase science identity among adolescent girls *J. Appl. Soc. Psychol.* **51** 667–82

[48] Chen C, Hardjo S, Sonnert G, Hui J and Sadler P 2023 The role of media in influencing students' STEM career interest *Intl J. STEM Educ.* **10** 56

[49] Kalender Z Y, Marshman E, Schunn C, Nokes-Malach T and Singh C 2019 Why female science, technology, engineering, and mathematics majors do not identify with physics: they do not think others see them that way *Phys. Rev. Phys. Educ. Res.* **15** 020148
Chen C, Sonnert G and Sadler P 2019 The effect of first high school science teacher's gender and gender matching on students' science identity in college *Sci. Educ.* **104** 75–99
Hazari Z, Sonnert G, Sadler P and Shanahan M C 2010 Connecting high school physics experiences, outcome expectations, physics identity, and physics career choice: a gender study *J. Res. Sci. Teach.* **47** 978–1003

[50] Hazari Z, Potvin G, Lock R, Lung F, Sonnert G and Sadler P 2013 Factors that affect the physical science career interest of female students: testing five common hypotheses *Phys. Rev. ST Phys. Educ. Res.* **9** 020115

[51] Nosek B, Banaji M and Greenwald A 2002 Math = male, me = female, therefore math not = me *J. Pers. Soc. Psychol.* **83** 44–59

[52] Logel C, Walton G M, Spencer S, Iseman E, von Hippel W and Bell A 2009 Interacting with sexist men triggers social identity threat among female engineers *J. Pers. Soc. Psych.* **96** 1089–103

[53] Cwik S and Singh C 2022 Self-efficacy and perceived recognition by peers, instructors, and teaching assistants in physics predict bioscience majors' science identity *PLoS One* **17** e0273621

[54] Hausmann L, Schofield J and Woods R 2007 Sense of belonging as a predictor of intentions to persist among African American and White first-year college students *Res. High. Educ.* **48** 803–39

[55] Stout J, Ito T, Finkelstein N and Pollock S 2013 How a gender gap in belonging contributes to the gender gap in physics participation *AIP Conf. Proc.* **1513** 402
Cwik S and Singh C 2022 Students' sense of belonging in introductory physics course for bioscience majors predicts their grades *Phys. Rev. Phys. Educ. Rev.* **18** 010139
Li Y and Singh C 2023 Sense of belonging is an important predictor of introductory physics students' academic performance *Phys. Rev. Phys. Educ. Res.* **19** 020137

[56] Hazari Z, Chari D, Potvin G and Brewe E 2020 The context dependence of physics identity: examining the role of performance/competence, recognition, interest, and sense of belonging for lower and upper female physics undergraduates *J. Res. Sci. Teach.* **57** 1583–607

[57] Shumba O and Glass L W 1994 Perceptions of coordinators of college freshman chemistry regarding selected goals and outcomes of high school chemistry *J. Res. Sci. Teach.* **31** 381–92

[58] Hazari Z, Tai R and Sadler P 2007 Gender differences in introductory university physics performance: the influence of high school physics preparation and affective factors *Sci. Educ.* **91** 847–76

[59] Stabiner K 2002 *All Girls: Single-Sex Education and Why it Matters* (New York: Riverhead)

[60] Pahlke E, Hyde J S and Allison C 2014 The effects of single-sex compared with co-educational schooling on students' performance and attitudes: a meta-analysis *Psychol. Bull.* **140** 1042–72
Smyth E 2010 Single-sex education: what does research tell us? *Rev. Fr. Pedagog.* **171** 47–55

[61] Kessels U and Hannover B 2008 When being a girl matters less: accessibility of gender-related self-knowledge in single-sex and coeducational classes and its impact on students' physics-related self-concept of ability *Br. J. Educ. Psychol.* **78** 283

[62] Sikora J 2014 Gender gap in school science: are single-sex schools important? *Sex Roles* **70** 411

[63] Institute of Physics 2003 Closing doors: exploring gender and subject choice in schools *IOP Guide* https://iop.org/about/publications/opening-doors (Accessed: 29 June 2024)

[64] Robinson W P and Gillibrand E 2004 Single-sex teaching and achievement in science *Int. J. Sci. Educ.* **26** 659–75

[65] Justman M and Mendez S 2018 Gendered choices of STEM subjects for matriculation are not driven by prior differences in mathematical achievement *Econ. Educ. Rev.* **64** 282–97

[66] Raviv A and Aflalo E 2023 Classroom discourse in single-sex physics classes: a case study *Eur. J. Sci. Math. Educ.* **11** 182–96

[67] Abraham J and Barker K 2020 Motivation and engagement with physics; a comparative study of females in single-sex and co-educational classrooms *Res. Sci. Educ.* **50** 2227–42

[68] Law H and Sikora J 2020 Do single-sex schools help Australians major in STEMM at university? *Sch. Effect. Sch. Improv.* **31** 605–27

[69] McGrayne S B 1993 *Nobel Prize Women in Science: Their Lives, Struggles and Momentous Discoveries* (Washington, DC: National Academies Press) p 287

[70] Kumagai J 1995 Do single-sex classes help girls succeed in physics? *Phys. Today* **48** 74

[71] Education resources *Institute of Physics* https://iop.org/education/ (Accessed: 3 June 2024)

[72] Potvin G *et al* 2023 Examining the effect of counternarratives about physics on women's physics career intentions *Phys. Rev. Phys. Educ. Res.* **19** 010126

[73] Dweck C 2007 *Mindset: The New Psychology of Success* (New York: Ballantine)

[74] Esparza J, Shumow L and Schmidt J 2014 Growth mindset of gifted seventh grade students in science *NCSSSMST J.* **19** 6–13

[75] Schmidt J, Shumow L and Kackar-Cam H 2015 Exploring teacher effects for mindset intervention outcomes in seventh-grade science classes *Middle Grades Res. J.* **10** 17–32

[76] Brown E 2014 'Growth mindset' may boost girls' confidence in STEM *Educ. Daily* **47** 1–2

[77] Macnamara B and Burgoyne A 2023 Do growth mindset interventions impact students' academic achievement? A systematic review and meta-analysis with recommendations for best practices *Psychol. Bull.* **149** 133–73

[78] Erkut S, Fields J, Sing R and Marx F 2002 Diversity in girls' experiences: feeling good about who you are *The Jossey-Bass Reader on Gender in Education* (San Francisco, CA: Jossey-Bass) p 500

[79] Hyde J 2005 The gender similarities hypothesis *Am. Psychol.* **60** 581–92

[80] AIP 2023 Percent of physics faculty members who are women, 2002–2022 *AIP Data Graphic* https://ww2.aip.org/statistics/percent-of-physics-faculty-members-who-are-women-2002-2022 (Accessed: 29 June 2024)

[81] Marks A, Dyer J, Monroe M and Butcher G 2015 Women in physics in the UK: update 2011–2014 *AIP Conf. Proc.* **1697** 060044

[82] Carrell S, Page M and West J 2010 Sex and science: how professor gender perpetuates the gender gap *Quart. J. Econ.* **125** 1101–44

[83] Sharma M and Bewes J 2011 Self-monitoring: confidence, academic achievement and gender differences in physics *J. Learn. Des.* **4** 1–13

[84] Fencl H and Scheel K 2005 Engaging students: an examination of the effects of teaching strategies on self-efficacy and course climate in a nonmajors physics course *J. Coll. Sci. Teach.* **35** 20–4
Fencl H and Scheel K 2004 Pedagogical approaches, contextual variables, and the development of student self-efficacy in undergraduate physics courses *AIP Conf. Proc.* **720** 173
Sawtelle V, Brewe E and Kramer L 2012 Exploring the relationship between self-efficacy and Q1 retention in introductory physics *J. Res. Sci. Teach.* **49** 1096–121
Cavallo A, Rozman M and Potter W 2004 Gender differences in learning constructs, shifts in learning constructs, and their relationship to course achievement in a structured inquiry, yearlong college physics course for life science majors *Sch. Sci. Math.* **104** 288–300

[85] Sawtelle V, Brewe E and Kramer L 2012 Exploring the relationship between self-efficacy and retention in introductory physics *J. Res. Sci. Teach* **49** 1096–121
Heilbronner N 2009 Pathways in STEM: factors affecting the retention and attrition of talented men and women from the STEM pipeline *PhD Dissertation* University of Connecticut, Storrs, ST

[86] Whitcomb K, Maries A and Singh C 2023 Progression in self-efficacy, interest, identity, sense of belonging, perceived recognition and effectiveness of peer interaction of physics majors and comparison with non-majors and PhD students *Res. Sci. Educ.* **52** 525–39

[87] Sundstrom M, Heim A, Park B and Holmes N 2022 Introductory physics students' recognition of strong peers: gender and racial or ethnic bias differ by course level and context *Phys. Rev. Phys. Educ. Res.* **18** 020148

[88] Bloodhart B, Balgopal M, Casper A M, Sample McMeeking L and Fischer E 2020 Outperforming yet undervalued: undergraduate women in STEM *PLoS One* **15** e0234685

[89] Rosenberg J, Holincheck N, Fernandez K, Dreyfus B, Wardere F, Stehle S and Butler T 2024 Role of mentorship, career conceptualization, and leadership in developing women's physics identity and belonging *Phys. Rev. Phys. Educ. Res.* **20** 010114

[90] Fine C 2010 *Delusions of Gender* (New York: Norton) p 51

[91] Fisher C, Brookes R and Thompson C 2022 I don't study physics anymore': a cross-institutional Australian study on factors impacting the persistence of undergraduate science students *Res. Sci. Educ.* **52** 1565–81

[92] Fisher C, Thompson C and Brookes R 2020 Gender differences in the Australian undergraduate STEM student experience: a systematic review *Higher Educ. Res. Dev.* **39** 1155–68

[93] Muehsler H E 2023 Differences in physics self-efficacy among personal and course-wider variables *Eur. J. Phys. Educ.* **14** 17–32

[94] Holmes N, Roll E and Bonn D A 2014 Participating in the physics lab: does gender matter? *Phys. Canada Special Issue* **70** 84–6
Guzzetti B and Williams W 1996 Gender, text, and discussion: examining intellectual safety in the science classroom *J. Res. Sci. Teach.* **33** 5–20

[95] Rosen D and Kelly A 2020 Epistemology, socialization, help seeking, and gender-based views in in-person and online, hands-on undergraduate physics laboratories *Phys. Rev. Phys. Educ. Res.* **16** 020116

[96] Archer L, Moote J, Francis B, DeWitt J and Yeomans L 2017 The 'exceptional' physics girl: a sociological analysis of multimethod data from young women aged 10–16 to explore gendered patterns of post-16 participation *Am. Educ. Res. J.* **54** 88–126

[97] Holmes N, Heath G, Hubenig K, Jeon S, Kalendar Z Y, Stump E and Sayre E 2022 Evaluating the role of student preference in physics lab group equity *Phys. Rev. Phys. Educ. Res.* **18** 010106

[98] Dew M, Hunt E, Perera V, Perry J, Ponti J and Loveridge A 2024 Group dynamics in inquiry-based labs: gender inequities and the efficacy of partner agreements *Phys. Rev. Phys. Educ. Res.* **20** 010121

[99] Doucette D 2021 Equity and introductory college physics labs *PhD Dissertation* University of Pittsburgh, PA

[100] Henderson R, Stewart J and Traxler A 2019 Partitioning the gender gap in physics conceptual inventories: force concept inventory, force and motion conceptual evaluation, and conceptual survey of electricity and magnetism *Phys. Rev. Phys. Educ. Res.* **15** 010131

[101] Nieberding M and Heckler A 2021 Patterns in assignment submission times: procrastination, gender, grades, and grade components *Phys. Rev. Phys. Educ. Res.* **17** 013106

[102] Simmons A and Heckler A 2020 Grades, grade component weighting, and demographic disparities in introductory physics *Phys. Rev. Phys. Educ. Res.* **16** 020125

[103] Dawkins H, Hedgeland H and Jordan S 2017 Impact of scaffolding and question structure on the gender gap *Phys. Rev. Phys. Educ. Res.* **13** 020117

[104] Kortemeyer G 2009 Gender differences in the use of an online homework system in an introductory physics course *Phys. Rev. ST Phys. Educ. Res.* **5** 010107

[105] Brewe E, Sawtelle V, Kramer L, O'Brien G, Rodriguez I and Pamela P 2010 Toward equity through participation in modeling instruction in introductory university physics *Phys. Rev. Phys. Educ. Res.* **6** 010106

[106] Kost-Smith L, Pollock S and Finkelstein N 2010 Gender disparities in second-semester college physics: the incremental effects of a 'smog of bias' *Phys. Rev. ST Phys. Educ. Res.* **6** 020112

[107] Wadsworth E and Clark B *Classroom Climate Workshop: Gender Equity* Teachers on Teaching: Purdue Video Workshops on College Teaching (West Lafayette, IN: Purdue University) (video)

[108] Rennie L J and Parker L H 1993 Assessment in physics: further explorations of the implications of item context *Aust. Sci. Teach. J.* **39** 28–32
Rennie L J and Parker L H 1991 Assessment of learning in science: the need to look closely at item characteristics *Aust. Sci. Teach. J.* **37** 56–9

[109] Hestenes D, Wells M and Swackhamer G 1992 Force concept inventory *Phys. Teach.* **30** 141

[110] McCullough L 2004 Gender, context, and physics assessment *J. Int. Women's Stud.* **5** 20–30
Majors T D W 2015 Conceptual physics differences by pedagogy and gender: questioning the deficit model *PhD Dissertation* Tennessee Technological University, Cookeville, TN

[111] Institute of Physics 2012 It's different for girls *Report* Institute of Physics p 21 https://iop.org/sites/default/files/2019-04/its-different-for-girls.pdf (Accessed: 29 June 2024)

[112] Gladys M J, Furst J E, Holdsworth J L and Dastoor P C 2023 Gender bias in first-year multiple-choice physics examinations *Phys. Rev. Phys. Educ. Res.* **19** 020109
Wilson K, Low D, Verdon M and Verdon A 2016 Differences in gender performance on competitive physics selection tests *Phys. Rev. Phys. Educ. Res.* **12** 020111

[113] Bug A, Hoshino-Brown E and Lui K 2011 Unconscious gender bias in the classroom *CSWP Gazette* **30** 12–13
Graves A, Hoshino-Brown E and Lui K 2016 Swimming against the tide: gender bias in the physics classroom *J. Women Minorities Sci. Eng.* **23** 15–36

[114] Fan Y *et al* 2019 Gender and cultural bias in student evalutions: why representation matters *PLoS One* **14** e0209749

[115] Potvin G and Hazari Z 2016 Student evaluations of physics teachers: on the stability and persistence of gender bias *Phys. Rev. Phys. Educ. Res.* **12** 020107

[116] Storage D, Horne Z, Cimpian A and Leslie S-J 2016 The frequency of 'brilliant' and 'genius' in teaching evaluations predicts the representation of women and African Americans across fields *PLoS One* **11** e0150194

[117] Malespina A, Schunn C and Singh C 2022 Whose ability and growth matter? Gender, mindset and performance in physics *Intl J. Sci. Educ.* **9** 28

[118] Korsunsky B 2000 Good odd days *Phys. Teach.* **38** 186–88
Lopac V, Tonejc A and Pecina P 2002 The past and present of physics in Croatia: gender differences in graduation statistics and textbook illustrations *AIP Conf. Proc.* **628** 149

[119] Larsen K 1995 Women in astronomy: inclusion in introductory textbooks *Am. J. Phys.* **63** 126

[120] Lawlor T and Niiler T 2020 Physics textbooks from 1960–2016: a history of gender and racial bias *Phys. Teach.* **58** 320–3

[121] UNESCO 2009 Promoting gender equality through textbooks *Programme and Meeting Document* ED.2009/WS/39 https://unesdoc.unesco.org/ark:/48223/pf0000158897_eng (Accessed: 29 June 2024)

[122] Donald A 2023 *Not Just for the Boys: Why We Need More Women in Science* (Oxford: Oxford University Press) p 133

[123] Seymour E and Hewitt N 1997 *Talking about Leaving* (Boulder, CO: Westview Press)

[124] Sadker M and Sadker D 1994 *Failing at Fairness* (New York: Touchstone) p 162

[125] Chappell B 2015 Nobel Laureate in hot water for 'trouble with girls' in labs *The Two-Way* http://npr.org/sections/thetwo-way/2015/06/10/413429407/nobel-laureate-in-hot-water-for-trouble-with-girls-in-labs (Accessed: 29 June 2024)

[126] Corneliussen S T 2014 Sexual harassment of scientists by scientists draws media attention *Phys. Today* 18 July
Straumsheim C 2015 'We all felt trapped' *Inside Higher Ed* 22 Jan www.insidehighered.com/news/2015/01/23/complainant-unprecedented-walter-lewin-sexual-harassment-case-comes-forward (Accessed: 29 June 2024)
Feltman R 2015 After years of sexually harassing students, superstar astronomer gets light warning *The Washington Post* 12 Oct www.washingtonpost.com/news/speaking-of-science/wp/2015/10/12/superstar-astronomer-sexually-harassed-students-for-years-according-to-investigation/ (Accessed: 29 June 2024)
Wilson R 2004 Louts in the lab *The Chronicle of Higher Education* **50** A7–9

[127] Aycock L, Hazari Z, Brewer E, Clancy K, Hodapp T and Goertzen R M 2019 Sexual harassment reported by undergraduate female physicists *Phys. Rev. Phys. Educ. Res.* **15** 010121

[128] Weale S 2014 New Cambridge students to attend compulsory sexual consent workshops *The Guardian* 19 Sept http://theguardian.com/education/2014/sep/19/new-cambridge-university-students-attend-compulsory-sexual-consent-workshops (Accessed: 27 January 2016)

[129] National Science Foundation 2016 The National Science Foundation (NSF) will not tolerate harassment at grantee institutions *NSF Press Release* http://nsf.gov/news/news_summ.jsp?cntn_id=137466 (Accessed: 29 June 2024)

[130] Harsh J, Maltese A and Tai R 2012 A perspective of gender differences in chemistry and physics undergraduate research experiences *J. Chem. Educ.* **89** 1364–70

[131] Conferences for undergraduate women and gender minorities *American Physical Society* https://aps.org/initiatives/inclusion/gender-inclusive/undergraduate-women-minorities (Accessed: 13 June 2024)

[132] Conference for Undergraduate Women and Non-Binary Physicists UK and Ireland (CUWiP+) *Institute of Physics* https://iop.org/physics-community/cuwip-uk-and-ireland (Accessed: 13 June 2024)

[133] Xie Y and Shauman K A 2003 *Women in Science: Career Processes and Outcomes* (Cambridge, MA: Harvard University Press) p 212

[134] Etzkowitz H, Kemelgor C and Uzzi B 2000 *Athena Unbound* (Cambridge: Cambridge University Press)
Davis C-S, Ginorio A, Hollenshead C, Lazarus B and Rayman P 1996 *The Equity Equation* (San Francisco, CA: Jossey-Bass)

[135] Etzkowitz H, Kemelgor C and Uzzi B 2000 *Athena Unbound* (Cambridge: Cambridge University Press) 79

[136] Hall L E 2007 *Who's Afraid of Marie Curie?* (Emeryville, CA: Seal) 134

[137] Kelsky K 2014 The 5 top traits of the worst advisors *The Professor Is In* http://theprofessorisin.com/2014/02/23/the-5-top-traits-of-the-worst-advisors/ (Accessed: 13 June 2024)

[138] Rigler K L, Bowlin L K, Sweat K, Watts S and Throne R 2017 Agency, socialization, and support: a critical review of doctoral student attrition *Proc. 3rd Int. Conf. on Doctoral Education* (Orlando, FL: University of Central Florida)

[139] Gu Y 2012 The influence of protégé–mentor relationship and social networks on women doctoral students' academic career aspirations in physical sciences and engineering *PhD Dissertation* University of California, Los Angeles, CA

[140] Verostek M, Miller C and Zwickl B 2024 Physics PhD student perspectives on the importance and difficulty of finding a research group *Phys. Rev. Phys. Educ. Res.* **20** 0101366

[141] Davis C-S, Ginorio A, Hollenshead C, Lazarus B and Rayman P 1996 *The Equity Equation* (San Francisco, CA: Jossey-Bass)
Grant C 2006 *Mentoring Success Strategies for Women in Science* ed P Pritchard (Burlington, MA: Academic)

[142] Etzkowitz H, Kemelgor C and Uzzi B 2000 *Athena Unbound* (Cambridge: Cambridge University Press) p 81

[143] Des Jardins J 2010 *The Madame Curie Complex: The Hidden History of Women in Science* (New York: The Feminist Press at CUNY)

[144] Hall R and Sandler B 1982 The classroom climate: a chilly one for women? *Report for the Project on the Status and Education of Women* (Washington, DC: Association of American Colleges)

[145] Sawtelle V, Brewe E and Kramer L 2012 Exploring the relationship between self-efficacy and retention in introductory physics *J. Res. Sci. Teach.* **49** 1096–121

[146] McAlister A, Lilly S and Chiu J 2021 Exploring factors that impact physical science doctoral student role identities through a multiple case study approach *Sci. Educ.* **106** 1501–34

[147] Project Juno *Institute of Physics* https://iop.org/about/IOP-diversity-inclusion/project-juno (Accessed: 13 June 2024)

[148] Climate site visits *American Physical Society* https://aps.org/initiatives/inclusion/climate/site-visits (Accessed: 13 June 2024)

[149] Sea Change *American Association for the Advancement of Science* https://seachange.aaas.org/

[150] Glanz J 1996 How not to pick a physicist? *Science* **274** 710–12

[151] Miller C, Zwickl B, Posselt J, Silvestrini R and Hodapp T 2019 Typical physics PhD admissions criteria limit access to underrepresented groups but fail to predict doctoral completion *Sci. Adv.* **5** eaat7550

[152] Verostek M, Miller C and Zwickl B 2021 Analyzing admissions metrics as predictors of graduate GPA and whether graduate GPA mediates PhD completion *Phys. Rev. Phys. Educ. Res.* **17** 020115

[153] Ralston C 2006 The making of a synchrotron geek *She's Such a Geek* ed A Newitz and C Anders (Emeryville, CA: Seal) p 78

[154] Gonsalves A 2014 'Physics and the girly girl—there is a contradiction somewhere': doctoral students' positioning around discourses of gender and competence in physics *Cult. Stud. Sci. Educ.* **9** 512

[155] Bad advisor horror stories *Reddit* https://reddit.com/r/GradSchool/comments/37ue2k/bad_advisor_horror_stories/ (Accessed: 13 June 2024)

IOP Publishing

Women and Physics (Second Edition)

Laura McCullough

Chapter 7

Careers

Until our scientific and technological workplace reflects our diversity, we
are not working to our potential as a nation.
—US Congresswoman Constance Morella [1]

Most of our lives are spent working, and many want a long and successful career.
Though each person's definition of success differs, to be successful requires many
pieces, and career-wise you need to be productive, make social connections, get good
advice, and (if you're a woman) overcome societal and cultural barriers. This
chapter looks into different parts of getting a job and advancing a career.

7.1 Getting the job

Once a woman has a degree or two in physics, she is usually ready to start looking
for a job. We see from the numbers (chapter 4) that the under-representation of
women continues in academia and industry. The 2022 data from the APS Statistical
Research Center shows women in about 20% of physics faculty jobs, though only
13% at the full professor rank [2].

The hiring process is almost certainly impacted by implicit bias, though to what
degree is an open question. As noted in chapter 3, unconscious biases about gender
and science mean that both men and women under-rate women's potential and over-
rate men's.

For jobs in the sciences, we don't know exactly what the impacts on total
employment numbers are, or the quality of any individual hiring process, though we
can make inferences from the data on implicit bias that there are real impacts on
attitudes and compensation. In terms of specific jobs or hires, a National Academy
of Sciences report in 2010 notes that 'data on the hiring process…are scant', partly
because the problems involved in different hiring practices among organizations and
institutions, and privacy issues regarding who has applied [3]. In their survey of SEM
departments (missing the Technology in STEM), they found that in physics, with

about 15% of available applicants (women with PhDs), there were on average 13% women in the pool for tenure-track positions. Not too bad. They also report that having a woman as committee chair or women on the committee increased the number of women in the applicant pool. Women were offered interviews at a higher percent than men, with interviews in physics being about 20% women. When interviewed, women were more likely than men to say that family-related reasons were part of why they accepted the job.

A 2005 experiment on biology, engineering, economics, and psychology faculty hiring showed a 2:1 preference for *women* for tenure-track positions [4]. Their analysis showed no differences between any of the four fields. They also tested several parameters beyond gender, including marriage status, having children, and taking parental leave in graduate school. It is worth noting that these researchers have a large corpus of work, all of which shows few or no gender differences when looking at particular aspects of STEM careers, although this does not comport with the findings of the majority of research in the field.

A more typical result from a study found that science faculty looking at identical-but-for-gender CVs for a lab manager preferred the male candidate, considering them more competent and hirable [5]—and offered them a higher starting salary. Physics faculty judging postdoc applicants showed a gender bias favoring males, also considering them more competent and hirable [6].

Unconscious bias may also interact with stereotyped behavior to compound impacts. In an experiment asking people to choose who to 'hire' for an arithmetic task that men and women perform equally well on, men were chosen more often when based (a) just on gender, (b) when candidates self-reported their expected performance, and (c) when candidates self-reported their past performance [7]. The only way women and men were equally valued was when the outside experimenter gave information on past performance. Of particular interest, the researchers noted that men were more likely to boast/overestimate their future performance, explaining the gap when self-reported performance was shared.

Besides your CV, another important issue for the job search is reference letters. The research here tends to not be field-specific, or, where it is, to focus on medical personnel. On the general side, letters for academics broadly showed a tendency in a 2009 study to describe women as more communal and less agentic than men [8]. An often-cited paper from medicine notes that women received more grindstone words (hard-working, dedicated) while men received more standout words (outstanding, amazing) [9].

There are very few studies on this topic specific to science. Geoscience postdoc reference letters for men were more likely to be 'excellent' than for women, though length did not differ [10]. A 2007 study for chemistry and biochemistry faculty positions noted that there were few gender differences except for male candidates receiving more standout adjectives [11]. (Chemistry has approximately 27% female assistant professors and 27% female postdocs.) One study of experimental particle physics and social science letters found few gender differences in either field [12].

It may be we have started to fix this bias! If you are interested in finding out if you use gendered words in your own letters of recommendation, search for a gender bias calculator such as Forth's program [13].

A quick note of warning: in 2023 a group looked at how large language models (LLMs, part of generative AI) did on writing letters for male and female candidates [14]. They found significant gender bias based on a short prompt. Part of the title of this research paper is 'Kelly is a warm person, Joseph is a role model'.

A rather alarming recent paper on US faculty found that a very small number of institutions is responsible for the undergraduate education of a large number of faculty [15]. Note that *undergraduate*. We already knew that the prestige of one's graduate institution had a disproportionate effect on faculty employment [16], particularly on women [17]. Now we see that it also plays a role at the undergraduate level.

7.2 Basic resources

In one of the earliest irrefutable pieces of evidence of bias in the sciences, Nancy Hopkins at MIT started a movement that ended up with a committee exploring gender bias in the School of Science. She took a tape measure to various labs to show that women had less lab space than men. And that was just a part of the issue. The committee found

> documented differences in salary in the recent past, in amount of nine-month salary paid from grants, in access to space, resources, and inclusion in positions of power and administrative responsibility within departments or within the broader MIT community. [18]

And it wasn't just at MIT [19].

The 2007 report by a US Committee 'Maximizing the potential of women in academic science and engineering' on bias in academic sciences and engineering found that women were experiencing barriers both structural and cultural at most parts of their career [20]. A later study in 2010 by the National Academy of Science found that in physics men had significantly more lab space than women, even when accounting for whether their work was experimental or theoretical [3]. Both men and women felt they received sufficient equipment, though women were less likely to feel they received enough clerical support. Women in physics were almost twice as likely to have a mentor, which is very gratifying! Electrical engineering was the same way. I would love to see another study like this done now.

More recently, an international study of the global gender gap in science was analyzed by discipline [21]. Women in physics reported fewer resources on average than men. And in a 2017 study by the American Institute of Physics, women earned lower salaries than men [22]. A study on over 1 million resumes in the US showed the widest gender gap in the sciences: women earn 87 cents to the men's dollar [23]. In the UK, men in science continue to make more than women [24]. Based on HESA data for 2022–23, women in physics make less than men among academic staff [25].

And while there were five women in the lower salary bracket, there were zero men. We still have a ways to go, but there is good evidence of positive movement.

7.3 Grant applications

To do science, you need resources. Whether it is a supercomputer or a supercollider, graduate students or lab managers, resources cost money. And most of that money comes from grant awards. Grants are a foundational part of physics. And as with most elements of physics, there is gender bias here too. The most famous study on gender and funding is Wenneras and Wold's 1997 study of the Swedish Medical Research Council, where they calculated that 'a female applicant had to be 2.5 times more productive than the average male applicant to receive the same competence score as he'. [26] This got a lot of press at the time. However, since then, the literature has shown mixed results.

At the broadest level, women in the US are just as likely as men to receive grants across all fields [27]. Women receive smaller awards, however, and fewer women apply than are eligible. This seems to hold across North America, Europe, New Zealand, and Australia [28]. Specific to the US National Science Foundation, grant success is equal for men and women across six directorates, including Mathematical and Physical Sciences, though with fewer submissions by women [29]. (An older study of NSF overall found women to be less likely to get an award and to get less money [30].) Physics applications to the Irish Research Council showed better success for women, with 23% of applications but 29% of awards [31]. In Canada, there are data for the Institutes of Health Research and the Social Science and Humanities Research Council, where women of color earned higher awards than men of color—in the humanities [32]. For the Canadian Institutes of Health Research, looking at the principal investigator (PI) as well as the science reduced the likelihood of women getting an award, while looking only at the science kept it balanced [33]. A study specific to Quebec noted that after the age of 38, women in natural sciences and engineering receive less funding [34]. In New Zealand, interviews with female academics noted several issues with receiving funding. The very title of the paper is revealing: 'Unless you are collaborating with a big name successful professor, you are unlikely to receive funding'. [35] This paper has a nice literature review of gender and funding.

In 2006 an Australian and German study of experimental physicists showed signs of the 'Matthew effect', in which previous success was key to getting new awards, leading to bias against early-career researchers [36]. This effect was also examined in a study of funding by the UK Research Council, along with noting that women researchers submit fewer proposals, leading to fewer awards [37]. In 2023 a study of the US National Institutes of Health funding showed women and people of color were significantly underrepresented as 'super PIs', who had multiple simultaneous grants [38].

So what can help? Double-blind reviews, for one. When the reviewer doesn't know the names on the grant application, more of the focus is on the project, and less on the investigator. Hubble Space Grant awards became significantly more gender balanced after instituting double-blind policies [39]. This helped out early-career scientists as well. Blind auditions for orchestras may have helped women (though not people of color) advance to prestigious groups [40], but only if the woman didn't wear high-heeled shoes that gave away her likely gender—a confounding factor that led them to put in carpeting after it was discovered.

Elsevier has been conducting large international studies on the general area of gender in science and innovation. In their 2024 report, they note that among a number of countries, women grant awardees now reach 37% [41]. Progress is being made. And I found one bit of very good news: Science Foundation Ireland saw an increase in female award holders after developing initiatives to mitigate gender bias [42]. It's always good to see that interventions can work.

Since grants are often a large part of retention and promotion decisions, particularly in research-oriented institutions, having a bias against women in the grant process means that women may be less likely to be retained and promoted. Cascade effects that can be compounded by other bias factors is a matter of great concern in the field.

7.4 Awards and honors

One of the highlights of any career is to receive an award. How do awards play out for women in physics? Let's start with our fellows—excuse me, our Fellows. Many professional organizations have the honor called Fellow. It represents significant accomplishments by a person and conveys great honor. And a woman can be a Fellow, despite how strange it sounds.

For the IOP Honorary Fellows, 28% of the 2022 award cohort was women, a significant improvement from ten years ago [43] (the IOP membership is approximately 22% women). The American Association of Physics Teachers Fellows included 23% women up to 2022 [44]. Data on APS awards from 1953 to 2017 show a 7% award rate for women [45].

Statements about desired diversity in awards are available for the APS [46], IOP [43], and the CAP [47]. The Canadian Association of Physics also requires selection committees to meet equity, diversity, and inclusion education standards [48].

Do we need to talk about the Physics Nobel Prize? Five women, 122 years, 225 laureates. Only two of those women before 2018. Let's just leave it there.

The IOP has done very well at looking in the mirror: they have intentionally examined gender bias in their awards [43]. In 2022, 32% of award winners were women—and they note that 19% of physics academic staff are women.

It's been well-known that science prizes have gender bias; it even has its own name: the Matilda effect [49], in an obvious nod to the Matthew effect mentioned above. Beyond simple gender bias favoring males, there are other forms of bias in award recognition. Women are more likely to earn teaching awards than research awards from physical science professional societies [50]. The same is true for service awards. The numbers of women earning awards from organizations can be inflated when women-only awards are included, in some cases quite significantly [50].

This may backfire though, by de-valuing women's scientific achievements. You got an award? Wow! Oh, it was one of *those* awards?

The selection committee also makes a difference. Looking at APS awards, Lincoln and colleagues found that the likelihood of a woman winning a physics award doubled for every woman added to the committee [51]. And more importantly, 'women are 65 percent less likely than men to win an award if the selection committee chair is a man, regardless of the number of women on the committee' [51].

7.5 Publishing

At this point, I'm sure you'll be shocked to hear that there is gender bias in publishing, right? The way it manifests is multi-fold. First, editors of scientific journals are more likely to be men. Second, peer reviewers are more likely to be men. Third, manuscripts authored by women are more likely to be rejected. Fourth, women are less likely to be cited.

The data on publication bias tend to be from studies across broad disciplines, and is also more common for the biological fields than the physical sciences, so much of what is described here is in those contexts.

Being a journal editor is a leadership position and is considered an honor. In the mathematical sciences, only 7.6% of journal editorship positions were held by women, despite women being 15% of tenure-stream faculty [52]. Examining the *Proceedings of the National Academy of Sciences* (PNAS), the first female editor was appointed in 1985, a molecular biologist [53]. In 2019, the Editorial Board was 24% female.

In physics, editors tend to be more senior, averaging 24 years of experience, more than other fields [54]. While the percentage of women physicists is under 20%, the percentage of women editors is even lower, around 16%. (A 4% difference doesn't sound bad, right ? But that is a drop of 20% from the available to the actual.)

In the only study on physics article submissions I could find, the Institute of Physics in 2018 examined its portfolio for diversity [55]. Women made up 27% of their submissions, and 22% of acceptances. This gave a 43% acceptance rate for men and 40% for women. Not too bad. They also looked at subfields and the lowest rate of female participation was in education and astrophysics journals. Interesting, since those fields tend to have more women [56, 57].

When a paper is submitted, it goes out for peer review. Who those reviewers are can make a difference in acceptance rates. In AGU journals, both authors and editors suggest or invite fewer women to review [58]. A shockingly low 22% of *Nature* reviewers in 2015 were women [59]. The Frontier series of journals had a lower rate of women in the peer-review process than the rate of women in the field would suggest [60].

Peer reviewing in publishing is similar to peer reviewing in grants: double-blind makes a difference [61]. Another option is transparent review, where readers can see reviewer reports, author responses, and editorial letters. The IOP Publishing house is the first of the physics journal publishers to institute both of these two types of reviews and has seen a positive effect on the regional diversity of its authors, though there are no differences between genders [62].

One small study in computer science found *single*-blind reviewing for conference papers increased the likelihood of accepting papers from top names and top institutions [63]. While not statistically significant, women were also less likely to be accepted.

In a large study of international physical science journals, physics had one of the highest gender gaps in number of papers published, citation gaps, and career length, although there was no gap in number of papers per year [64]. Women are at least as productive annually, but with shorter overall careers. Another large study examining journal data estimates that physics (using arXiv data) had approximately 15% female authors, with a rate of change of approximately 0.25% per year [65]. Specific to contemporary physics (1995–2020), a citation gap of about 4% in favor of men was found, with interesting variations by subfield. [66]

For astrophysics and astronomy, women's authorship in prestigious journals has gone up smoothly over the last 50 years, reaching about 20% in 2017 [67]. In theoretical physics, the rate was around 10% with very small increases by decade.

The American Geophysical Union's journals show a higher rate of acceptance of women's papers, but invitations to referee are still slanted towards men [58]. Interestingly, women were more likely to decline an invitation to peer review. Possibly a result of woman having a higher service load than men.

Within the JSTOR corpus (heavy on the biological sciences), women were less likely to have a higher authorship position (first or last, usually) and were only 27% of authors overall [68]. The gender gap in engineering authorship is high: men account for about 80% of all scientific production [69].

A study of medical journals found that women were significantly less likely to have invited commentaries [70]. Similar results were found for invited essays for *Nature* and *Science* [71]. And among *PNAS* and *Science* journals, women were less likely to comment on research—men's research in particular [72].

Citations of women's work are lower than men's [73]. In astronomy, women's papers were 10% less likely to be cited than men's [74], and similar results are found in engineering [69]. And men are more likely to cite themselves than women overall, including in mathematics [75].

7.6 Sexual harassment

While there is a lack of research on some topics that women in science face, sexual harassment isn't one of them. The bad side: how much it happens. The good side: increased awareness. The best place to start is the 2018 US National Academies study 'Sexual harassment of women: climate, culture, and consequences in academic sciences, engineering, and medicine' [76]. The basic news? It happens. A lot. This report was launched at the same time that the #MeToo movement was spreading across social media, bringing the topic into further prominence.

Not all harassment is the same. Most research looks at three types: sexist gender harassment, sexual gender harassment, and sexual harassment. Sexist gender harassment is most common, exemplified by the 'women can't do physics' type of comment. Sexual gender harassment is not unwanted attention, but more along the lines of posting pictures of scantily-clad women or calling all women 'sluts'. Sexual harassment is unwanted sexual attention, whether it be come-ons (hostile environment) or demands for sex in return for something (*quid pro quo*).

Overall, academia is a hotbed for harassment: 58% of female academics reported experiencing harassment [76]. Government and the private sector were less bad, at 43% and 46%, respectively. Factors that support an environment where harassment happens include: a perceived tolerance for harassment, male-dominated workplaces, a hierarchical power structure, symbolic compliance with Title IX and Title VII (check boxes instead of actual policies), and uninformed leadership. Fear of retaliation is a big problem with low reporting rates, as is not wanting to be drawn into a big (and likely public) investigation. Unfortunately, if no one reports, the behavior can continue for years or decades.

In physics specifically, we only have a little data on how prevalent this problem is. Undergraduate women in the US provided a startling and dismaying statistic: three-quarters of them had experienced some form of harassment [77]. Seventy-four percent. It's amazing that women get even 20% of physics degrees! That 74% matches with data from a study on field-work scientists, where 64% reported being a victim of sexual harassment, and 20% for sexual assault [78]. At McMurdo station in Antarctica, 59% of women have experienced sexual harassment or assault [79]. Female physics graduate students are targets for both microaggressions as well as harassment [80], and 30% of female astronomy graduate students felt unsafe at work, compared to 2% for men [81]. And women of color had it much worse—50% felt unsafe at work.

While physics is not a field science, we do have large collaborations, including places where women may be working alone at night, such as at CERN. CERN recently created a landing page for ethics, including a specific anti-harassment policy [82]. I suspect their gender diversity policy was prioritized after Alessandro Strumia's anti-women comments garnered world-wide attention [83]. He was disinvited from his position at CERN and has been sanctioned.

After the National Academies report was released, a new coalition was formed by several organizations: the Societies Consortium on Sexual Harassment in STEM, with a goal of helping scientific organizations reduce harassment [84]. Many organizations and professional groups have some form of a code of conduct or ethics statement, explicitly forbidding harassing behavior [85]. Some have made harassment part of their definition of scientific misconduct [86], which allows for stronger sanctions. Others have statements that behavior which falls outside of the code of conduct or scientific misconduct can result in loss of privileges, including attending conferences and even revoking honors [87]. In physics and astronomy, the two most prominent incidents have been regarding Geoff Marcy (expelled from the National Academy of Science) [88] and Lawrence Krauss (placed on administrative leave, then agreed to retire) [89]. The Wellcome Trust and the NSF require disclosure of harassment behavior of principal investigators.

If you have experienced harassment, or are afraid you will, you are spending time and energy on things that aren't physics. It affects mental health, well-being, and performance [90]. As well as making you wonder if you belong [77].

7.7 Imposter phenomenon

If you don't feel that you belong in a field, or if you don't feel your own identity matches with a physics identity, you are likely to have some emotional and psychological turmoil. One of the ways that this can manifest is imposterism or imposter phenomenon. (Originally called imposter syndrome [91], the name has been changed to get away from the suggestion that it is a disease. What a surprise— an issue with women and it's a 'syndrome'. Although men get it too.)

Imposter phenomenon (IP) is the feeling that even though you have succeeded, you really don't belong or deserve your success, and that you will be outed as a fraud. You got lucky, or people gave you a break, or *something* that wasn't due to your performance and ability is the true reason for your success. As comedian and actor Tina Fey put it: 'The beauty of the imposter syndrome is you vacillate between extreme egomania, and a complete feeling of "I'm a fraud! Oh god, they're onto me! I'm a fraud!"' [92] These feelings can lead to thoughts of leaving the field: US astronomy grad students who experienced it were more likely to consider leaving, though they did not actually leave at higher rates because of IP [93]. The environment you're in matters: undergraduate women who experienced harassment were also more likely to experience IP [77].

There isn't a lot of research out there specifically looking at IP and physics. But the 'brilliance' perception shows up here too: fields with a perceived requirement of brilliance (such as physics) have more women—especially women of color—who experience IP [94]. I'm beginning to think that getting rid of that particular stigma for physics will do more for the diversity in the field than many other interventions. Personally, whenever I get the response of 'oh, you must be smart' when I share what I do, I say 'you don't have to be smart to do physics; you just need a good teacher'. That's my own bias showing.

It's not clear how IP affects actual performance; it's something that high-achievers experience. Undergraduates in the US had a correlation between higher high school grades and higher feelings of IP [95]. College students of color experience more mental health problems from IP than from racial stress [96]. There is an argument that, as with stereotype threat [97], feelings of IP use up vital mental resources, leading to problems with performance [98]. US postdocs in STEM who had feelings of imposter phenomenon did not pursue new things, had a hard time networking, and lacked confidence in communicating: all things important to careers [99].

There isn't much work on IP in the UK context, but one study of marginalized STEMM undergraduates argues that IP is not actually a personality phenomenon but something that the culture imposes or produces on those who are not behaving to cultural norms [100]. Similarly, US doctoral students in STEM felt that institutional culture and sexism were contributing factors to IP [101].

By this point, you've probably noticed that most of the research is on students and higher ed more generally. As noted earlier, it is quite hard to find general career studies that are focused on women in science.

A survey on women leaders in STEM found over 60% of them had experienced imposter phenomenon in either their STEM or their leadership roles [102]. Unfortunately, IP has a positive relationship with work–family conflict [103]. A double whammy.

7.8 Mentors and sponsors

It is always a comfort when you know you can go to someone for advice, recommendations, or commiseration. Mentors can provide these things. Among US physicists, mentors were mentioned as one reason for career success, with women being more likely to mention them [104]. Astronomy graduate students with good mentors were less likely to think about leaving the field [93], and female mentors for undergraduate researchers had positive effects on their protegees [105]. For undergraduates in science, mentors had the biggest effect on career choice [106]. And 66% of female STEM leaders identified mentors as an assistance in their path to leadership [102].

Mentorship is unevenly apportioned, however. One study found that in physics, men were more than three times as likely as women to have worked with an 'eminent' mentor [107]. And women don't need a female mentor, just a good one. See chapter 6 for more research on this and advisor relationships.

Sponsorship is a different thing from mentorship [108]. A sponsor isn't someone you go to for emotional or psychological help, or for advice in navigating culture. The job of a sponsor is to help advance your career, often by recommending you for honors and opportunities [109]. A sponsor doesn't even have to have a strong connection to you. While there isn't any research on sponsorship in physics, in general it is understood to help promote diversity [110]. Sponsorship programs, if carefully designed, could help increase diversity in STEM [111].

7.9 Networking and advancement

Along with mentors and sponsors, it can be important to grow your network. Who you know may help or hinder you in terms of opportunities, especially for important committees and invited talks. Women's networks during graduate school in the physical sciences and engineering helped them in socialization as well as career decision making [112]. Conferences are a key part of building a network, and can help with reducing isolation: one physics grad student laughed that 'she never knew that there were so many women working in physics' [112]. In this study, women noted that conference activities focused on women were very helpful for providing a sense of community and a supportive network.

In the Earth and space sciences, women had more men in their networks than men had women, and women had more male coauthors than the proportion of men in the field would suggest [113]. Peer networks have been one of the most common interventions among NSF ADVANCE grant programs for improving gender equity in STEM, and have helped to promote agentic actions for women's advancement [114].

Among academic scientists, women are more likely to have leadership positions than men, but more in disciplinary leadership positions as opposed to university or research group leadership [115]. In other words, the less prestigious ones. And the more women in your network, the less likely you are to have a leadership position. Probably because 'they do not generate as much social capital for those possibly wanting a leadership position' [115].

Athene Donald notes that your network is one way you can receive help with 'amplification' [116]. When you propose an idea and someone else co-opts it, an ally or network member can redirect the credit back to you: 'I see you are building on Natasha's idea', or 'Thanks for reminding us of what Talia said', are examples of this. Donald also reminds us that setbacks and failure are not only part of science, but we can use them to learn [117].

In the Global Survey of Physicists, men were more likely to have had certain advancement opportunities. They were more likely to have acted as a manager, given an invited talk, and been a journal editor [118]. Networks can help start changing this.

7.10 Leadership

Given the low number of women in physics, is there a representative number of women in physics leadership? It's hard to know. Limited research is available on the topic.

In Canada, a Supreme Court decision paved the way for an increase in the number of women holding NSERC Agency Canada Research Chairs, from 17% in 2016 to 28% in 2020 [119]. Good legislation helps: in several countries women now hold more leadership positions on boards and in legislatures [120]. But quotas alone can help *and* hinder. Women and men may be less interested in leadership if they think a woman was promoted based only on quotas [121]. This can be ameliorated by intentional sharing of a woman's qualifications.

A snapshot study in 2019 of US department chairs found that in the top physics departments only 15% of chairs were women, and in a random sampling of departments the number dropped to 10% [122].

As a principal investigator, or research group leader, women may face extra barriers from their research group. A study of chemistry research groups found that graduate students had gender expectations of their female supervisor, causing problems when they behaved according to gender norms (empathic, nurturing) and also when they behaved according to 'leader' norms (critical, impersonal) [123]. The leadership double bind for women has been known for a long time—try *Through the Labyrinth* for more on this [124].

In an interesting study of leadership, research, and service roles, a name switch (Cathy or Charlie) study found that engineering respondents were more likely to recommend Cathy as co-chair [125]. The authors posit 'bias correction', where people are aware of past bias and attempts to correct or be neutral. Something to be hopeful about!

7.11 Other issues

There are plenty of other issues that affect women's careers in physics. Among the things I did not touch on were collaboration preferences, tokenism, communication styles, negotiation, and choosing non-academic careers (anathema to many academics). If I were to include everything I would like to look at on this topic, it would quickly balloon what is supposed to be a chapter into an entire new book of its own.

References

[1] Morella C 2000 Land of plenty: diversity as America's competitive edge in science, engineering and technology *Report* cawmset0409 Congressional Commission on the Advancement of Women and Minorities in Science, Engineering and Technology Development https://www.nsf.gov/pubs/2000/cawmset0409/cawmset_0409.pdf

[2] Porter A M and Tyler J 2024 The state of the academic workforce in physics and astronomy departments, 2000–2022 *AIP Report* https://ww2.aip.org/statistics/the-state-of-the-academic-workforce-in-physics-and-astronomy-departments-2000-2022 (Accessed: 29 June 2024)

[3] National Academy of Sciences 2010 *Gender Differences at Critical Transitions in the Careers of Science, Engineering, and Mathematics Faculty* (Washington, DC: The National Academies Press) 42

[4] Williams W and Ceci S 2015 National hiring experiments reveal 2:1 faculty preference for women on STEM tenure track *Proc. Natl Acad. Sci.* **112** 5360–5
Ceci 2018 Women in academic science: experimental findings from hiring studies *Educ. Psych.* **53** 22–41

[5] Moss-Racusin C, Dovidio J, Brescoll V, Graham M and Handelsman J 2012 Science faculty's subtle gender biases favor male students *Proc. Natl Acad. Sci.* **109** 16474–9

[6] Eaton A, Saunders J, Jacobson R and West K 2019 How gender and race stereotypes impact the advancement of scholars in STEM *Sex Roles* **82** 127–41

[7] Reuben E, Sapienza P and Zingales L 2014 How stereotypes impair women's careers in science *Proc. Natl Acad. Sci.* **111** 4403–8

[8] Madera J, Hebl M and Martin R 2009 Gender and letters of recommendation for academia: agentic and communal differences *J. Appl. Psychol.* **94** 1591–9

[9] Trix F and Psenka C 2003 Exploring the color of glass: letters of recommendation for female and male medical faculty *Discourse Soc.* **14** 191–220

[10] Dutt K, Pfaff D, Bernstein A, Dillar J and Block C 2016 Gender differences in recommendation letters for postdoctoral fellowships in geoscience *Nat. Geosci.* **9** 805–8

[11] Schmader T, Whitehead J and Wysocki V 2007 A linguistic comparison of letters of recommendation for male and female chemistry and biochemistry job applicants *Sex Roles* **57** 509–14

[12] Bernstein R, Macy M, Williams W, Cameron C, Williams-Ceci S C and Ceci S 2022 Assessing gender bias in particle physics and social science recommendations for academic jobs *Soc. Sci.* **11** 74

[13] Gender bias calculator *Tom Forth* https://tomforth.co.uk/genderbias/ (Accessed: 28 June 2024)

[14] Wan Y, Pu G, Sun J, Garimella A, Change K W and Peng N 2023 Kelly is a warm person, Joseph is a role model; gender biases in LLM-generated reference letters arXiv: 2310.09219

[15] Grunspacn D, Komperda R, Offerdahl E, Abraham A, Etebari S, Maas S, Roberts J, Ghafoor S and Brownell S 2024 Importance of undergraduate institution prestige in physics faculty hiring networks *Phys. Rev. Phys. Educ. Res.* **20** 010144

[16] Wapman K, Zhang S, Clauset A and Larremore D 2022 Quantifying hierarchy and dynamics in US faculty hiring and retention *Nature* **610** 120

[17] Clauset A, Arbesman S and Larremore D B 2015 Systematic inequality and hierarchy in faculty hiring networks *Sci. Adv.* **1** e140005

[18] MIT 1999 A study on the status of women faculty in science at MIT *MIT Faculty Newsletter* **vol 11** March https://web.mit.edu/FNL/women/women.html (Accessed: 29 June 2024) 13

[19] Tilghman S 2004 Ensuring the future participation of women in science, mathematics, and engineering *The Markey Scholars Conference Proceedings* (Washington, DC: National Academies Press)

[20] Institute of Medicine 2007 *Beyond Bias and Barriers: Fulfilling the Potential of Women in Academic Science and Engineering* (Washington, DC: The National Academies Press)

[21] Ivie R and White S 2023 The global survey of scientists: results for physics *Presentation* American Institute of Physics https://gender-equality-in-science.org/wp-content/uploads/2023/06/RIvieSurvey.pdf (Accessed: 24 June 2024)

[22] Feder T 2017 Salaries for female physics faculty trail those for male colleagues *Phys. Today* **70** 24–6

[23] Westfall C 2024 Inequality doubles in gender pay gap for women, new survey shows *Forbes* 14 May https://forbes.com/sites/chriswestfall/2024/05/14/inequality-doubles-in-gender-pay-gap-for-women-new-survey-shows/ (Accessed: 20 May 2024)

[24] New Scientist 2024 The New Scientist 2024 global talent trends and insights *Report* https://newscientist.com/nsj/article/global-talent-trends-and-insights (Accessed: 23 June 2024)

[25] HESA 2024 What are their salaries? https://hesa.ac.uk/data-and-analysis/staff/salaries (Accessed: 23 June 2024)

[26] Wenneras C and Wold A 1997 Nepotism and sexism in peer review *Nature* **387** 342

[27] Schmaling K and Gallo S 2023 Gender differences in peer reviewed grant applications, awards, and amounts: a systematic review and meta-analysis *Res. Integ. Peer Rev.* **8** 2

[28] Marsh H W, Jayasinghe U W and Bond N W 2011 Gender differences in peer reviews of grant applications: a substantive-methodological synergy in support of the null hypothesis *J. Informetrics* **5** 167–80
Leberman S, Eames B and Barnett S 2016 Unless you are collaborating with a big name successful professor, you are unlikely to receive funding *Gender Educ.* **28** 644–61

[29] Rissler L, Hale K, Joffe N and Caruso N 2020 Gender differences in grant submissions across science and engineering fields at the NSF *BioScience* **70** 814–20

[30] Hosek S, Cox A, Ghosh-Dastidar B, Kofner A, Ramphal N, Scott J and Berry S 2005 *Gender Differences in Major Federal External Grant Programs* (Santa Monica, CA: RAND Corporation)

[31] Kavanagh Y, Gilheany S, McLoughlin E, Byrne M, Arredondo M, Poppenhaeger K and Maher N 2019 How successful are women in physics in Ireland? *AIP Conf. Proc.* **2109** 050021

[32] Nakhaie R, Lippert R and Cukarski D 2023 Granting inequities: racialization and gender differences in Social Science and Humanities Research Council of Canada's grant amount for research elites *Can. Ethnic Stud.* **55** 25–49

[33] Witteman H, Hendrick M, Straus S and Tannenbaum C 2018 Female grant applicants are equally successful when peer reviewers assess the science, but not when they assess the scientists bioRxiv: 10.1101/232868 (Accessed: 21 June 2024)

[34] Lariviere V, Vignola-Gagne E, Villeneuve C, Gelinas P and Gingras Y 2011 Sex differences in research funding, productivity, and impact: and analysis of Quebec University professors *Scientometrics* **87** 483–98

[35] Leberman S, Eames B and Barnett S 2016 Unless you are collaborating with a big name successful professor, you are unlikely to receive funding *Gender Educ.* **28** 644–61

[36] Laudel G 2006 The 'quality myth': promoting and hindering conditions for acquiring research funds *Higher Educ.* **52** 375–403

[37] Viner N, Powell P and Green R 2004 Institutionalized biases in the awards of research grants: a preliminary analysis revisitng the principle of accumulative advantage *Res. Pol.* **33** 443–54

[38] Nguyen M, Chaudhry S, Desai M, Dzirasa K, Cavazos J and Boatright D 2023 Gender, racial, and ethnic inequities in receipt of multiple National Institutes of Health research program grants *JAMA Network Open* **6** e230855

[39] Chawla D 2021 Record number of first-time observers get Hubble telescope time *Nature News* 25 Nov

[40] Goldin C and Rouse C 2000 Orchestrating impartiality: the impact of 'blind' auditions on female musicians *Am. Econ. Rev.* **90** 715–41
Gelman A 2019 Did blind orchestra auditions really benefit women? *Statistical Modeling, Causal Inference, and Social Science* https://statmodeling.stat.columbia.edu/2019/05/11/did-blind-orchestra-auditions-really-benefit-women/ (Accessed: 21 June 2024)
Dworkin A 2021 Rethinking blind auditions *Symphony* https://symphony.org/features/rethinking-blind-auditions/ (Accessed: 21 June 2024)

[41] Elsevier 2024 Progress towards gender equality in research and innovation—2024 review *Report* https://elsevier.com/en-xs/insights/gender-and-diversity-in-research (Accessed: 21 June 2024)

[42] Fritch R, McIntosh A, Stokes N and Boland M 2019 Practitioners' perspectives: a funder's experience of addressing gender balance in its portfolio of awards *Interdisc. Sci. Rev.* **44** 192–203

[43] Supporting diversity and excellence in IOP Awards *Institute of Physics* https://iop.org/about/awards/supporting-diversity-excellence-iop-awards (Accessed: 18 June 2024)

[44] McCullough L 2022 Women's leadership in physics education *Poster at AAPT meeting in Grand Rapids MI, July 2022* https://lauramccphd.com/index.php/home/research/selected-presentations/

[45] Private communication, Scott Franklin, 26 March 2017

[46] APS Fellowship *American Physical Association* https://aps.org/funding-recognition/fellowship/aps-fellowship

[47] Canadian Association of Physicists 2022 CAP's Equity, Diversity and Inclusion Statement https://cap.ca/edi/

[48] Fellows Program Details *Canadian Association of Physicists* https://cap.ca/programs/recognitions/post-graduate-physicists/fellows-program-intro/details/

[49] Rossiter M 1993 The Matthew Matilda effect in science *Soc. Stud. Sci.* **23** 325–41

[50] Lincoln A, Pincus S, Bandows Koster J and Leboy P S 2012 The Matilda effect in science: awards and prizes in the US, 1990s and 2000s *Soc. Stud. Sci.* **42** 307–20

[51] Lincoln A E, Pincus S and Schick V 2009 Evaluating science or evaluating gender? *Am. Phys. Soc. News* **18** 8

[52] Topaz C M and Sen S 2016 Gender representation on journal editorial boards in the mathematical sciences *PLoS One* **11** e0161357

[53] Berenbaum M 2019 Speaking of gender bias *Proc. Natl Acad. Sci.* **116** 8086–8

[54] Liu F, Holme P, Chiesa M, AlShebli B and Rahwan T 2023 Gender inequality and self-publication are common among academic editors *Nat. Hum. Behav.* **7** 353–64

[55] IOP Publishing 2018 Diversity and inclusion in peer review at IOP Publishing *Report* https://ioppublishing.org/wp-content/uploads/2018/09/J-VAR-BK-0818-PRW-report-final.pdf (Accessed: 24 June 2024)

[56] Tyler J 2024 Degrees earned in the physical sciences and engineering fields *AIP Data Graphic* https://ww2.aip.org/statistics/physics-engineering-degrees-earned (Accessed: 29 June 2024)

[57] Barthelemy R, Van Dusen B and Henderson C 2015 Physics education research: a research subfield of physics with gender parity *Phys. Rev. Phys. Educ. Res.* **11** 020107

[58] Lerback J and Hanson B 2017 Journals invite too few women to referee *Nature* **541** 455–7

[59] [no author] 2017 Gender imbalance in science journals is still pervasive *Nature* **541** 435–6

[60] Helmer M, Schottdorf M, Neef A and Battaglia D 2017 Gender bias in scholarly review *eLife* **6** e21718

[61] Kern-Goldberger A, James R, Berghella V and Miller E 2022 The impact of double-blind peer review on gender bias in scientific publishing: a systematic review *Am. J. Obst. Gyn.* **227** 43–50

Mulligan A, Hall L and Raphael E 2012 Peer review in a changing world: an international study measuring the attitudes of researchers *J. Amer. Soc. Info. Sci. Tech.* **64** 132–61

Budden A, Tregenze T, Aarssen L, Koricheva J, Leimu R and Lortie C 2008 Double-blind review favours increased representation of female authors *Trends Ecol. Evol.* **23** 4–6

[62] Holst F, Eggleton K and Harris S 2022 Transparency versus anonymity: which is better to eliminate bias in peer review? *Insights: UKSG J.* **35** 1–6

[63] Tomkins A, Zhang M and Heavlin W 2017 Single versus double blind reviewing at WSDM 2017 arXiv: 1702.00502

[64] Huang J, Gates A, Sinatra R and Barabasi A-L 2020 Historical comparison of gender inequality in scientific careers across countries and disciplines *Proc. Natl Acad. Sci.* **117** 4609–16

[65] Holman L, Stuart-Fox S and Hauser C 2018 The gender gap in science: how long until women are equally represented? *PLoS Biol.* **16** e2004956

[66] Teich E *et al* 2022 Citation inequity and gendered citation practices in contemporary physics *Nat. Phys.* **18** 1161–70

[67] Mihaljevic H and Santamaria L 2020 Authorship in top-ranked mathematical and physical journals: role of gender on self-perceptions and bibliographic evidence *Quant. Sci. Stud.* **1** 1468–92

[68] West J, Jacquet J, King M, Correll S and Bergstrom C 2013 The role of gender in scholarly authorship *PLoS One* **8** e66212

[69] Ghiasi G, Lariviere V and Sugimoto C 2015 On the compliance of women engineers with a gendered scientific system *PLoS One* **10** e0145931

[70] Thomas E, Jayabalasingham B, Collins T, Geertzen J, Bui C and Dominici F 2019 Gender disparities in invited commentary authorship in 2459 medical journals *JAMA Network Open* **2** e1913682

[71] Conley D and Stadmark J 2012 Gender matters: a call to commission more women writers *Nature* **488** 590

[72] Wu C, Fuller S, Shi Z and Wilkes R 2020 The gender gap in commenting: women are less likely than men to comment on (men's) published research *PLoS One* **15** e0230043

[73] Lariviere V, Ni C, Gingras Y, Cronin B and Sugimoto C R 2013 Bibliometrics: global gender disparities in science *Nature* **504** 211–3
Elsevier 2024 Progress towards gender equality in research and innovation—2024 review *Report* https://elsevier.com/en-xs/insights/gender-and-diversity-in-research (Accessed: 21 June 2024)

[74] Caplar N, Tacchella S and Birrer S 2017 Quantitative evaluation of gender bias in astronomical publications from citation counts *Nat. Astron.* **1** 0141

[75] King M, Bergstrom C, Correll S, Jacquet J and West J 2017 Men set their own cites high: gender and self-citation across fields and over time *Socius* **3** 1–22

[76] National Academies of Sciences, Engineering, and Medicine 2018 *Sexual Harassment of Women: Climate, Culture, and Consequences in Academic Sciences, Engineering, and Medicine* (Washington, DC: National Academies Press)

[77] Aycock L, Hazari Z, Brewer E, Clancy K, Hodapp T and Goertzen R M 2019 Sexual harassment reported by undergraduate female physicists *Phys. Rev. Phys. Educ. Res.* **15** 010121

[78] Clancy K, Nelson R, Rutherford J and Hinde K 2014 Survey of academic field experiences (SAFE): trainees report harassment and assault *PLoS One* **9** e102172

[79] US National Science Foundation 2022 Results from the US Antarctic Program's sexual assault and harassment needs assessment *Report* https://nsf.gov/news/news_summ.jsp?cntn_id=305782&org=OPP (Accessed: 25 June 2024)

[80] Barthelemy R, McCormick M and Henderson C 2016 Gender discrimination in physics and astronomy: graduate student experiences of sexism and gender microaggressions *Phys. Rev. Phys. Educ. Res.* **12** 020119

[81] Clancy K, Lee K, Rodgers E and Richey C 2017 Double jeopardy in astronomy and planetary science: women of color face greater risks of gendered and racial harassment *J. Geophys. Res. Planets* **122** 1610–23

[82] Ethics at CERN https://hr.web.cern.ch/ethics

[83] Alessandro Strumia *Wikipedia* https://en.wikipedia.org/wiki/Alessandro_Strumia

[84] Societies Consortium on Sexual Harassment in STEMM https://societiesconsortium.com/

[85] For example, Ethics guidelines *American Physical Society* https://aps.org/about/governance/policies-procedures/ethics
AAS code of ethics *American Astronomical Society* https://aas.org/policies/ethics

[86] Scientific ethics at AGU https://agu.org/learn-about-agu/about-agu/ethics

[87] Revocation process *AAAS* https://aaas.org/programs/fellows/revocation-process
Revocation policy *American Physical Society* https://aps.org/about/governance/policies-procedures/revocation

[88] Kaiser J 2021 Astronomer Geoff Marcy booted from National Academy of Sciences in wake of sexual harassment *Science Insider* 27 May

[89] Mangan 2018 3 revelations from the Lawrence Krauss sexual-harassment report *Chronicle Higher Educ.* **65** 23 Oct https://www.chronicle.com/article/3-revelations-from-the-lawrence-krauss-sexual-harassment-report/

[90] Swim J, Hyers L, Cohen L and Ferguson M 2001 Everyday sexism: evidence for its incidence, nature, and psychological impact from three daily diary studies *J. Soc. Issues* **57** 31–53

Dardenne B, Dumont M and Bollier T 2007 Insidious dangers of benevolent sexism: consequences for women's performance *J. Pers. Soc. Psych.* **93** 764–79

[91] Clance P R and Imes S A 1978 The imposter phenomenon in high-achieving women: dynamics and therapeutic intervention *Psychother. Theor. Res. Pract.* **15** 241–7

[92] Pringle C 2010 Tina Fey—from spoofer to movie stardom *Independent* 19 Mar https://independent.co.uk/arts-entertainment/films/features/tina-fey-from-spoofer-to-movie-stardom-1923552.html (Accessed: 25 June 2024)

[93] Ivie R, White S and Chu R 2016 Women's and men's career choices in astronomy and astrophysics *Phys. Rev. Phys. Educ. Res.* **12** 020109

[94] Muradoglu M, Horne Z, Hammond M, Leslie S J and Cimpian A 2022 Women—particularly underrepresented minority women—and early-career academics feel like impostors in fields that value brilliance *J. Ed. Psych.* **114** 1086–100

[95] King J and Cooley E 1995 Achievement orientation and the imposter phenomenon among college students *Cont. Educ. Psych.* **20** 304–12

[96] Cokley K, McClain S, Enciso A and Martinez M 2013 An examination of the impact of minority status stress and impostor feelings on the mental health of diverse ethnic minority college students *J. Multicult. Couns. Dev.* **41** 66–122

[97] Pennington C, Heim D, Levy A and Larkin D 2016 Twenty years of stereotype threat research: a review of psychological mediators *PLoS One* **11** e0146487

[98] Hudson S and Gonzalez-Gomez H 2021 Can imposters thrive at work? The imposter phenomenon's role in work and career outcomes *J. Voc. Behav.* **128** 103601

[99] Chakraverty D 2020 The impostor phenomenon among postdoctoral trainees in STEM: a US-based mixed-methods study *Int. J. Dr. Stud.* **15** 329–52

[100] Murray O, Chiu Y T, Wong B and Horsburgh J 2023 Deindividualising imposter syndrome: imposter work among marginalized STEMM undergraduates in the UK *Sociology* **57** 749–66

[101] Bano S and O'Shea C 2023 Factors contributing to imposter phenomenon in doctoral students: a US-based qualitative study *Int. J. Dr. Stud.* **18** 251–69

[102] McCullough L 2020 Barriers and assistance for female leaders in academic STEM in the US *Educ. Sci.* **10** 264

[103] Gullifor D, Gardner W, Karam E, Noghani F and Cogliser C 2023 The imposter phenomenon at work: a systematic evidence-based review, conceptual development, and agenda for future research *J. Organ. Behav.* **45** 234–51

[104] Porter A M 2019 Physics PhDs ten years later: success factors and barriers in career paths *AIP Report* https://aip.org/statistics/reports/physics-phds-ten-years-later-success-factors-and-barriers-career-paths (Accessed: 13 June 2024)

[105] Moghe S, Baumgart K, Shaffer J J and Carlson K A 2021 Female mentors positive contribute to undergraduate STEM research experiences *PLoS One* **16** e0260646

[106] Downing R, Crosby F and Blake-Beard S 2005 The perceived importance of developmental relationship on women undergraduates' pursuit of science *Psych. Women Quart.* **29** 419–26

[107] Grant L and Ward K 2004 Structures of academic science careers and the progress of women *American Sociological Association Conference (San Francisco)*

[108] Cao J and Yang Y 2013 What are mentoring and sponsoring and how do they impact organizations? *Article* ILR School, Cornell University, Ithaca, NY https://ecommons.cornell.edu/items/19846299-1c7a-4b2c-b41e-aa4f8b3c183d/full (Accessed: 18 June 2024)

[109] Donald A 2023 *Not Just for the Boys* (Oxford: Oxford University Press)

[110] Mentorship and sponsorship are keys to unlocking the next generation of talent *Catalyst* 6 Feb https://catalyst.org/2024/02/06/mentorship-and-sponsorship-are-keys-to-unlocking-the-next-generation-of-talent/ (Accessed: 25 June 2024)

[111] Huston W, Cranfield C, Forbes S and Leigh A 2018 A sponsorship action plan for increasing diversity in STEMM *Ecol. Evol.* **9** 2340–5

[112] Gu Y 2012 The influence of protégé–mentor relationships and social networks on women doctoral students' academic career aspiration in physical sciences and engineering *PhD Dissertation* University of California, Los Angeles 86

[113] Hanson B, Wooden P and Lerback J 2020 Age, gender, and international author networks in the Earth and space sciences: implications for addressing implicit bias *Earth Space Sci.* **7** 1–11

[114] O'Meara K A and Stromquist N 2015 Faculty peer networks: role and relevance in advancing agency and gender equity *Gender Educ.* **27** 338–58

[115] Parker M and Welch M 2013 Professional networks, science ability, and gender determinants of three types of leadership in academic science and engineering *Lead. Quart.* **24** 344

[116] Donald A 2023 *Not Just for the Boys* (Oxford: Oxford University Press) p 224

[117] Donald A 2018 Nothing is wasted—turning negative experiences into positive life lessons *TED Talk* https://ted.com/talks/athene_donald_nothing_is_wasted_turning_negative_experiences_into_positive_life_lessons (Accessed: 28 June 2024)

[118] Porter A M and Ivie R 2019 *Women in Physics and Astronomy, 2019* (College Park, MD: AIP Statistical Research Center)

[119] Rangan C *et al* 2023 Status of gender equity in physics in Canada (2017–2020) *AIP Conf. Proc.* **3040** 050007

[120] Sojo V, Wood R, Wood S and Wheeler M 2016 Reporting requirements, targets, and quotas for women in leadership *Lead. Quart.* **27** 519–36

[121] Nater C, Heilman M and Sczesny S 2022 Footsteps I would like to follow? How gender quotas affect the acceptance of women leaders as role models and inspirations for leadership *Eur. J. Soc. Psych.* **53** 129–46

[122] McCullough L 2019 Proportions of women in STEM leadership in the academy in the USA *Educ. Sci.* **10** 1

[123] Hirschfield L 2014 She's not good with crying': the effect of gender expectations on graduate students' assessments of their principal investigators *Gender Educ.* **26** 601–17

[124] Eagly A and Carli L 2007 *Through the Labyrinth: The Truth About How Women Become Leaders* (Boston, MA: Harvard Business Review Press)

[125] Judson E, Ross L, Hjelmstad K, Krause S, Ankeny C, Culbertson R and Middleton J 2017 Examination of implicit gender biases among engineering faculty when assigning leadership, research, and service roles *Proc. of the ASEE Annual Conf. & Expositions 2017* pp 11440–53

Chapter 8

Work-life balance

[I]f you budget your time for work, you must budget your time for play as that is essential for every one. [...] Your physical condition rests on your mental condition and on your spiritual attitude toward life.

—Eleanor Roosevelt [1]

There is a strong stereotype that a career in science means giving up ... things. Children. Spouse. Vacations. The picture of the scientist working 12 or 14 hour days in the lab is a particularly harmful bit of cultural baggage [2]. But as with most stereotypes, there is truth hiding behind the picture. Postdocs in chemistry and physics spoke of tensions between family and science, and women postdocs were more likely than men to prioritize family over work [3]. What are the benefits and barriers for women in physics regarding family and home life ?

8.1 To have and to hold: marriage

Internationally, approximately 60% of physicists are married, according to a large study [4]. There is a higher proportion of married physicists in developing countries compared to developed countries. Women in developed countries felt their marriage had a positive effect on their career, unlike women from developing countries.

Beyond the typical benefits that may accrue from having a partner or children, are there other benefits for women in physics? In the study of postdocs mentioned at the beginning of this chapter, both women and men noted that family can be a support to a science career. Whether it is rides to work, taking a greater share of the housework, or simply being a shoulder to lean on, a partner can ease the load [3]. And in the international study, women spoke of having someone to talk to, discussing physics, and encouragement to continue on in physics.

I know very well the value of a supportive spouse. He drove me to graduate school each day, did the housework, and helped with typing when a medical issue left my hands too weak for the task. After my PhD graduation defense, my supervisor congratulated *him* first. Twenty-five years on, my husband (an author himself) did a quick read-through on this book before I turned it in.

Societal expectations can cause marriages to be more fractious, with women being expected to take care of the household and children. 'If I come home and continue work, then housework just doesn't get done. If I come home and do housework, then I do not produce as much for work as [my husband] does.' [4]

When a physicist publishes a paper, they must choose how to list themselves as an author. Does a woman include her full name, often making it clear that she is female? If she does, she may set herself up for potential bias on the part of reviewers or those who might choose to cite the article, whether the bias is explicit or implicit (see chapter 7). Or she may choose to limit herself to a first initial [5]. While this may subject her personally to less bias, it removes an important cultural reminder that *women do physics*, which may make things more difficult for women who follow her. This is one of many issues and hard choices that men in the same position are typically spared. Some journals use initials for everyone; a nice way to make this a moot point.

Another publishing dilemma that is primarily a woman's issue comes from the fact that many women change their last name upon marriage or divorce. When you've changed your name, how do you let people know that the Jane Doe who has published six condensed matter papers is the same Jane Smith that is now submitting another condensed matter manuscript? How do you let grant reviewers know that your bibliography has ten papers, not just two? There is very little research on how this affects female authors [6]. Transgender men and women have a similar issue regarding name changes plus the burden of outing themselves should they choose to request changes [7]. One proposed option is to use ORCID to push name changes out to journals invisibly. (If you are a researcher and don't have your ORCID number yet, go do that! [8].)

8.2 The two-body problem/dual-career couples

The two-body problem—the issue of prioritizing where to live, which opportunities to pursue, and other important decisions necessitated by having a two-career household—is also a particular difficulty for women. Having a working spouse is a different experience for men and women in heterosexual relationships. A woman is more likely to be a 'trailing spouse' and pause or end her career in order to promote her husband's [9]. An older survey found that female physicists are more likely to be married to another physicist than male physicists are [10]. The international study of physicists noted that many physicists reported being married to scientists or other physicists [4].

More recent research shows that among scientists in general, female scientists are more likely to be married to other scientists [11], making this a much more common problem for women in physics than for men, though there can be advantages to marrying another scientist: 'In my department, the most successful women are married to men who are also in the department' [5]. Spousal relationships can help women and men: 'Generations of women scientists were made into permanent research associates through marriage to other scientists' [12].

Historically there have been several famous physicist couples [13], proof that it is possible to succeed in such a relationship. While this can be a positive, it can also be a sign of deeply set issues of discrimination or bias, forcing women pursuing science in hostile environments or times into collaborations they might not have needed if their work were allowed to be evaluated solely on its own merits.

While it is illegal to discriminate based on marital status in most of the Anglophone counties (and a bad idea even if not illegal!), it is still likely a part of hiring decisions. A study of married candidates being considered for faculty positions found that women's marital status was discussed but men's was not [14].

While companies and universities sometimes have spousal hiring programs [15], they are only useful in limited situations and often provide only temporary employment to the spouse. And be aware of what you may give up in gaining a spousal hire: 'Realize that this goal will very likely, unless you are some kind of super-star, take many other negotiable elements of your offer off the table, such as substantially higher salary, research funds, etc' [16]. And even if you manage it, there are still burdens, particularly on the trailing spouse [17]. The difficulties are even greater for those in same-sex marriages [18]. There is an association in the US offering support to institutions looking at dual-career issues [19], and exposure to dual-career science couples can mitigate stereotypes that a science career is less family-friendly than other fields [20]. It can also create tensions in situations where one spouse is granted tenure and the other is not, or other such split decisions.

When two careers conflict, who gives up something? Among postdocs, 31% of married women were willing to make concessions to accommodate their spouse's career, compared to 21% of the men [21]. Among the same sample, 30% of men but only 15% of women expected their spouse to make concessions. These issues may become more prominent as the Millennial generation is more focused on work-life balance [22].

8.3 To breed or not to breed: children

In the US, the total fertility rate is 1.8—the average number of births a woman would have if all women lived through childbearing age [23]. In New Zealand and Australia the number is similar, as is Canada. Ireland and the UK are slightly lower, at 1.7 and 1.6. In the US, about 53% of women ages 15–50 have children [24], and about half of UK women have children [25]. Not everyone wants children, but certainly a majority of women do. Internationally, among a large sample of physicists, 43% have children [4]. Among women over 45, however, 79% have children. Interestingly, many women in the study (presumably younger) stated that they decided not to have children.

An interesting difference for male and female scientists is that having a child typically slows a woman's career progression, while it can have no effect [4] or a small positive effect [26] on a man's career. The research in this area is very mixed, however, especially when disaggregated by race, or across countries [27]. It is very difficult for women to have their cake and eat it too [28].

A 2018 international study of physicists asked about parenting, and men were over three times as likely as women to say that becoming a parent did not significantly affect their career [29]. Women were more likely to say they spent less time at work and their career slowed. A bonus though—women were more likely than men to say they became more productive and efficient at work!

Family-friendly policies such as stopping the tenure clock can help, but many women do not take advantage of stopping the tenure clock because of entirely reasonable fears of repercussions: 'I saw the two other women with young children get punished on reviews for not getting enough published even though they "had time off and had more time to write". I wasn't going to risk it' [30]. The impact of having a child is big enough that some women decide to have fewer children than they wanted, in order to keep their career going [31]. The decision to have fewer or no children can even have recruitment effects: during an 8th grade girls' science day a young girl asked how many of the present female scientists had kids. 'Not a single one of us did …. You could see the smiles on the girl's [sic] faces just dissolve' [32].

Our society still views motherhood as a key component to women's fulfillment in life. Even at the end of the twentieth century, it was believed that in order to truly succeed, a woman had to achieve domestic perfection as well as professional accomplishments to be considered a complete person: 'All women scientists should marry, rear children, cook, and clean in order to achieve fulfillment, to be a complete woman' [33]. This quote from Nobel-winning medical physicist Rosalyn Yalow (1921–2011) demonstrates that it's not just men who set unreasonable expectations of women. And it is part of why women don't enter, or leave, physics [34]. A meta-analysis of parenting for women of color in STEM is being finalized by Winterrowd; initial results are available [27]. She notes that 'There may still be a 'parenting penalty' or a 'gender penalty' but without interaction between the two.' This suggests some progress.

Even the best of us have experienced barriers—the amazing Millie Dresselhaus had trouble after giving birth to her fourth child. 'When she had her fourth kid', recalls a colleague, 'she brought him to work the day after he was born. She was there around noon or 1 o'clock with the baby in tow. But because Lincoln Lab was a government lab, you either had to have clearance or have a badge. They wouldn't let the kid in. She was furious! I didn't see her angry that often, but I saw her angry that day.' [35] (By the way, have you seen the joyous GM advertisement featuring Dresselhaus? [36] I watch it regularly.)

When women in science achieve greatness, the attention they receive is often heavily focused on their gender and family status. Upon being awarded a Nobel Prize in Physics in 1963, Maria Goeppert Mayer's success was announced 'S.D. MOTHER WINS NOBEL PHYSICS PRIZE' in a San Diego newspaper [37]. Mayer was interviewed for *McCall's* magazine (which remarked on the dress she wore to the Nobel ceremony), and *Science Digest* noted her beauty and how good a wife she was [38].

8.4 Housework and childcare

One of the international studies of physicists [39] asked about housework and expectations in the home. Regardless of country (and cultural gender roles) female physicists do more housework than their male colleagues. This pattern is true for the general population, even now [40]. Spending more time with housework obviously means less time for research and work.

Another issue is that of providing care for parents at the same time as providing care for children or grandchildren—being 'sandwiched' between generations. Estimates range from 30% to 60% of middle-aged people providing care both up and down [41]. And women are more likely to be the carers than men [42]. The data on this are across society broadly, excepting as regards COVID below.

8.5 Career breaks and re-entry

A database search for 'physics, women, career re-entry' produced zero results. Replacing 'physics' with 'science' produced six results, none of which were for the sciences. All I can find in this area are programs to support women looking to return to the workforce. The Wellcome Trust, APS, NSF, AWIS, and others have grants or scholarships for (mostly) women.

A search on career breaks gave me two conference papers, neither of which was specifically research, and two more articles from 1982 and 1996. In a nice display of irony, the article from 1982 is about the 'obsolescence of knowledge'. Indeed.

So we don't know how many physics women do take career breaks, or how many do not manage to re-enter the profession. There is a lot of advice out there, and a lot of articles about particular programs to support returning women, but no data.

This topic is one of the reasons I prefer the analogy of a highway [43] to a leaky pipeline [44]—there are on-ramps as well as off-ramps.

8.6 COVID

When the world went into lockdown because of the COVID pandemic, many households were turned upside-down. Childcare became a bigger part of the day for parents, and housework also gained prominence. Stress levels during the pandemic were worse for women and people staying with multiple children [45]. Readers who have gotten this far through the book probably do not find this surprising.

Some of the effects for women in STEM were blurred boundaries between work and home, less effective work time, increased 'sandwich care' (caring for children and elders at the same time), and reduced self-care [46]. This National Academies report also looked into the differences between tenure-track and non-tenure-track women faculty.

Not only did COVID disrupt family life, it disrupted scientific life. Lockdowns meant not getting to the lab or the classroom. So, what did a lot of scientists do? Pulled out those manuscripts and polished them up! Of course, given the differential effect of COVID on women, it is disappointing but not surprising that women submitted fewer papers than men did, especially for more junior women [47].

In a study of how NSF and NIH PIs in the US handled the lockdown, most PIs were very concerned about the health and safety of their research group members, as well as the health and safety of their family [48]. Interestingly, women placed greater priority than men on four things: making the right decision for personnel, how personnel felt about conducting on-site research, providing emotional support to personnel, and concerns about financial impact on personnel. In another study, female STEM professors noted concerns about their students: 'If the lab remains closed to standard research much longer, many students' mental health will deteriorate as they feel they are not progressing in their research.' [49]

Women's higher proportion of housework and childcare may even have become worse when a household lost working hours [50]. Research productivity measured by publications showed that female researchers took a heavy hit: a decrease in first authorship and last authorship, and bigger decreases for mid-career women and women from countries with lower gender-equality and stronger COVID restrictions [51]. For academics in STEM + Medicine, having a child under five years old reduced the number of peer reviews and funding panels completed [52], as well as a larger reduction in research time [53]. For women it was a harder hit, again because of the primary responsibility for childcare. The NAS report found that only 9% of their sample shared childcare equally with their spouse [54].

One study of female STEM academics found the parenting issue to be especially hard: 'I am exhausted, frustrated, stressed and unable to get done what is typically expected of me, while I see childless people enjoying this time and excelling.' [49] Personally, as someone who doesn't have children, I certainly didn't *enjoy* lockdown, but definitely had an easier time than my colleagues who were parents.

In an ironic coda, women may have been the biggest driver in economic recovery since the pandemic [55].

8.7 Mental health

One of the few silver linings of the pandemic was that it brought to light a lot of things that had been invisible, mental health being one of the most important.

College students in general have been struggling with mental health issues [56], exacerbated by the pandemic [57]. Physics students are no different [58]. And, unsurprisingly, it was worse for women. In STEM, mental health may be lower than in non-STEM fields: female engineering students had lower mental health scores than their male peers [59], although a recent report notes that women in male-dominated STEM fields have less anxiety and depression than women in gender-balanced STEM fields [60]. So there's some hope. Or maybe it's just confusion.

Graduate school is typically a very hard time in one's life, and mental health for graduate students is typically not great. A study at UC Berkeley noted over 40% of students in the physical sciences scored as depressed [61]. Black Women graduate students in computer science reported that 'onlyness' had impacts on their mental health, citing breakdowns and depression [62].

Female postdocs across countries reported mental health issues, and they often were connected with science identity: 'If you spend all your time doing something then that becomes sort of part of your identity. So then if things aren't going well in that aspect then it's harder to be positive about other things in your life.' [63]

In their book *Burnout*, the Nagoski sisters note that 20%–30% of teachers and professors in America have moderately high to high levels of burnout. They talk about the 'helping professions' as being particularly prone to burnout [64]. (I can recommend this book for genuinely *useful* suggestions for dealing with burnout, particularly for women.)

When a woman gets paid less than a male peer, she ends up having more depression and anxiety [65]. Less money AND more depression? Hardly seems fair. And for women, depression has a positive correlation with greater intent to leave a STEM job [66]. Burnout has been particularly high since the pandemic went endemic, and my inbox filled up with advice on avoiding burnout. If only I had time to read those articles [67].

Being a woman in a patriarchal society has always meant tensions between career and home life. For women in physics, we need to include the additional stress of a male-dominated career in the equation. It is not an easy path to follow, and this is one place where change feels particularly slow.

References

[1] Roosevelt E 2019 *It's Up to the Women* (New York: Bold Type) pp 54–5

[2] Cech E and Blair-Loy M 2022 *Misconceiving Merit: Paradoxes of Excellence and Devotion in Academic Science and Engineering* (Chicago, IL: University of Chicago Press)

[3] Wyss V and Tai R 2010 Conflicts between graduate study in science and family life *Coll. St. J.* **44** 475–91

[4] Ivie R and Guo S 2006 Women physicists speak again *AIP Report* www.aip.org/statistics/reports/women-physicists-speak-again (Accessed: 29 June 2024) p 13

[5] Swanson N 2004 *Penetrating the Tungsten Barrier* (Delhi: Imprint) p C-4

[6] Pellack L and Kappmeyer L 2011 The ripple effect of women's name changes in indexing, citation, and authority control *J. Am. Soc. Inf. Sci. Technol.* **62** 440–8

[7] Gaskins L C and McClain C R 2021 Visible name changes promote inequity for transgender researchers *PLoS Biol.* **19** e3001104

[8] ORCID https://orcid.org/

[9] Amcoff J and Niedomysl T 2015 Is the tied returnee male or female? The trailing spouse thesis reconsidered *Popul. Space Place* **21** 872–81

[10] McNeil L and Sher M 1999 Dual-science-career couples: survey results http://www.physics.wm.edu/~sher/survey.pdf (Accessed: 29 June 2024)

[11] Xie Y and Shauman K A 2003 *Women in Science: Career Processes and Outcomes* (Cambridge, MA: Harvard University Press)

[12] Gornick V 2009 Women in Science: Then and Now (New York: The Feminist Press) p 70

[13] Thomas W 2010 Married physicist couples *Ether Wave Propaganda* https://etherwave.wordpress.com/2010/05/05/married-physicist-couples/ (Accessed: 29 June 2024)

[14] Rivera L 2017 When two bodies are (not) a problem: gender and relationship status discrimination in academic hiring *Am. Soc. Rev.* **82** 111101138

[15] Feder T 2007 Workshop aims to double number of women in physics *Phys. Today* **60** 35–6

[16] Kelsky K 2011 Negotiating the spousal hire *The Professor Is In* 15 Dec https://theprofessor-isin.com/2011/12/15/negotiating-spousal-hire/ (Accessed: 29 June 2024)

[17] Anon 2014 Spousal hire realities *Inside Higher Ed* 3 Jun www.insidehighered.com/advice/2014/06/04/essay-what-its-be-spousal-hire-faculty-job (Accessed: 29 June 2024)
McNulty Y 2012 'Being dumped in to sink or swim': an empirical study of organizational support for the trailing spouse *Human Res. Dev. Int.* **15** 417–34

[18] Careless E and Mizzi R 2015 Reconstructing careers, shifting realities: understanding the difficulties facing trailing spouses in higher education *Can. J. Educ. Admin. Policy* 166 5 Mar

[19] Kibel L S 2013 166 Beyond the 'two-body' problem: recruitment with dual-career couples support *New England Board of Higher Education* 10 Jun https://nebhe.org/thejournal/beyond-the-two-body-problem-increasing-recruitment-roi-with-dual-career-couples-support/

[20] Tan-Wilson A and Stamp N 2015 College students' views of work-life balance in STEM research careers: addressing negative preconceptions *CBE-Life Sci. Educ.* **14** 1–13

[21] Martinez E, Botos J, Dohoney K, Geiman T, Kolla S, Olivera A, Qiu Y, Rayasam G V, Stavreva D and Cohen-Fix O 2007 Falling off the academic bandwagon *EMBO Rep.* **8** 977–81

[22] Tan-Wilson A and Stamp N 2015 College students' views of work–life balance in STEM research careers: addressing negative preconceptions *CBE—Life Sci. Educ* **14** 1–13

[23] CIA Country comparisons—total fertility rate *The World Factbook* www.cia.gov/the-world-factbook/field/total-fertility-rate/country-comparison (Accessed: 12 June 2024)

[24] The United States Census Bureau 2023 Fertility of women in the United States: 2020, table 2 https://census.gov/data/tables/2020/demo/fertility/women-fertility.html#par_list_58 (Accessed: 29 June 2024)

[25] Office for National Statistics 2022 Childbearing for women born in different years, England and Wales: 2020 *Census 2021* figure 2 https://ons.gov.uk/peoplepopulationandcommunity/birthsdeathsandmarriages/conceptionandfertilityrates/bulletins/childbearingforwomenbornindifferentyearsenglandandwales/2020 (Accessed: 29 June 2024)

[26] Mason M A and Goulden M 2004 Marriage and baby blues: redefining gender equity in the academy *Ann. Am. Acad. Polit. Soc. Sci.* **596** 86–103

[27] Winterrowd E 2019 Academic motherhood: a mixed-method review of the 'child penalty' for women of color in STEM *ARC Network* (video) https://equityinstem.org/blog/academic-motherhood (Accessed: 12 June 2024)

[28] Williams W and Ceci S 2012 When scientists choose motherhood *Am. Sci.* **100** 138–45

[29] Ivie R and White S 2023 The global survey of scientists: results for physics *Presentation* American Institute of Physics https://gender-equality-in-science.org/wp-content/uploads/2023/06/RIvieSurvey.pdf (Accessed: 24 June 2024)

[30] Frasch K, Stacy A, Mason M A, Page-Medrich S and Goulden M 2009 The devil is in the details *Establishing the Family Friendly Campus: Models for Effective Practice* ed J Lester and M Sallee (Sterling, VA: Stylus) p 91

[31] Mason M A, Wolfinger N H and Goulden M 2002 *Do Babies Matter? Gender and Family in the Ivory Tower* (New Brunswick, NJ: Rutgers University Press)

[32] Swanson N 2004 *Penetrating the Tungsten Barrier* (Delhi: Imprint Books) p 10

[33] Brown D 2011 Nobel winner Rosalyn Yalow dies at 89 *The Washington Post* 2 Jun www.washingtonpost.com/local/obituaries/nobel-winner-rosalyn-yalow-dies-at-89/2011/06/02/AGwgMdHH_story_1.html (Accessed: 29 June 2024) (paywall)

[34] Eren E 2022 Never the right time: maternity planning alongside a science career in academia *J. Gend. Stud.* **31** 136–47

Thebaud S and Taylor C 2021 The spectre of motherhood: culture and the production of gendered career aspirations in science and engineering *Gender Soc.* **35** 395–421

[35] Anderson M 2015 Mildred Dresselhaus: Queen of Carbon *IEEE Spectr.* **52** 50–4

[36] GE #BalanceTheEquation Campaign: 'What If scientists were celebrities?' *Meaningful Impact* (video) https://youtube.com/watch?v=PdhzZ56D4Kc (Accessed: 29 June 2024)

[37] This month in physics history *American Physical Society* www.aps.org/publications/apsnews/201312/physicshistory.cfm (Accessed: 29 June 2024)

[38] Rossiter M 1995 *Women Scientists in America: Before Affirmative Action* (Baltimore, MA: Johns Hopkins Press)

[39] Ivie R and Tesfaye C 2012 Women in physics: a tale of limits *Phys. Today* **65** 47–50

[40] Pew Research Center 2015 Raising kids and running a household: how working parents share the load https://www.pewresearch.org/social-trends/2015/11/04/raising-kids-and-running-a-household-how-working-parents-share-the-load/ (Accessed: 29 June 2024)
Lyonette C 2015 Sharing the load? Partners' relative earnings and the division of domestic labour *Work Employ. Soc.* **29** 23–40

[41] Vlachantoni A, Evandrou M, Falkingham J and Gomez-Leon M 2020 Caught in the middle in mid-life: provision of care across multiple generations *Ageing Soc.* **40** 1490–510
McGarrigle C and Kenny R A 2013 *Profile of the Sandwich Generation and Intergenerational Transfers in Ireland* (Dublin: The Irish Longitudinal Study on Ageing)

[42] Evandrou M, Falkingham J, Gomez-Leon M and Vlachantoni A 2018 Intergenerational flow of support between parents and adult children in Britain *Ageing Soc.* **38** 321–51
Friedman E, Park S and Wiemers E 2017 New estimates of the sandwich generation in the 2013 panel study of income dynamics *The Gerontologist* **57** 191–6

[43] Anderson-Rowland M 2009 The engineering highway: a new metaphor especially appropriate for women *WEPAN 2009 National Conf. Proc.* https://journals.psu.edu/wepan/article/view/58580

[44] Marshall J 2008 Escape from the pipeline: women using physics outside academia *Phys. Teach.* **46** 20–4

[45] Kowal M *et al* 2020 Who is the most stressed during the COVID-19 pandemic? Data from 26 countries and areas *App. Psych.: Health and Well-being* **12** 946–66

[46] National Academies of Sciences, Engineering, and Medicine 2021 *The Impact of COVID-19 on the Careers of Women in Academic Sciences, Engineering, and Medicine* (Washington, DC: The National Academies Press)

[47] Squazzoni F, Bravo G, Grimaldo F, Garcia-Cost D, Farjam M and Mehmani B 2021 Gender gap in journal submissions and peer review during the first wave of the COVID-19 pandemic. A study on 2329 Elsevier journals *PLoS One* **16** e0257919

[48] Antes A, McIntosh T, Solomon Cargill S, Bruton S and Baldwin K 2023 Principal investigators' priorities and perceived barriers and facilitators when making decisions about conducting essential research in the COVID-19 pandemic *Sci. Eng. Ethics* **29** 8

[49] Dunn M, Gregor M, Robinson S, Ferrer A, Campbell-Halfaker D and Marin-Fernandez J 2022 Academia during the time of COVID-19: examining the voices of untenured female professors in STEM *J. Career Assess.* **30** 573–89 p 582-3

[50] Zamberlan A, Gioaching F and Gritti D 2021 Work less, help out more? The persistence of gender inequality in housework and childcare during UK COVID-19 *Res. Soc. Strat. Mobil.* **73** 100583

[51] Kwon E, Yun J and Kang J 2023 The effect of the COVID-19 pandemic on gendered research productivity and its correlates *J. Informetrics* **17** 101380

[52] Krukowski R, Reshma J and Cardel M 2021 Academic productivity differences by gender and child age in science, technology, engineering, mathematics, and medicine faculty during the COVID-19 pandemic *J. Women's Health* **30** 341–7

[53] Myers K *et al* 2020 Unequal effects of the COVID-19 pandemic on scientists *Nat. Hum. Beh.* **4** 880–3

[54] National Academies of Sciences, Engineering, and Medicine 2021 The impact of COVID-19 on the careers of women in academic sciences, engineering, and medicine (Washington, DC: The National Academies Press)

[55] Bauer L and Wang S Y 2023 Prime-age women are going above and beyond in the labor market recovery *The Hamilton Project* 30 Aug https://hamiltonproject.org/publication/post/prime-age-women-are-going-above-and-beyond-in-the-labor-market-recovery/ (Accessed: 26 June 2024)

[56] National Healthy Minds study https://healthymindsnetwork.org/hms/ (Accessed: 29 June 2024)

[57] Fruehwirth J C, Biswas S and Perreira K M 2021 The COVID-19 pandemic and mental health of first-year college students: examining the effect of COVID-19 stressors using longitudinal data *PLoS One* **16** e0247999

[58] Wright M 2023 We can't ignore the pandemic fallout in our physics classes *Phys. Teach.* **61** 326–7

[59] Deziel M, Olawo D, Truchon L and Golab L 2013 Analyzing the mental health of engineering students using classification and regression *Educational Data Mining Conference*

[60] Reidy D and Wood L 2024 The mental health of undergraduate women majoring in STEM *J. Am. Coll. Health* 1–6

[61] UC Berkeley Graduate Assembly 2014 Graduate student happiness and well-being *Report* https://gradresources.org/wp-content/uploads/2015/09/wellbeingreport_2014-17.pdf

[62] Joshi A P, Shavers M C, Spencer B M, Artis S and LeSure S 2024 Exploring the impact of 'onlyness' among Black women doctoral students in computer science and engineering *J. Div. High. Educ.* https://doi.org/10.1037/dhe0000583

[63] Ysseldyk R, Greenaway K, Hassinger E, Zutrauen S, Lintz J, Bhatia M, Frye M, Starkenburg E and Tai V 2019 A leak in the academic pipeline: identity and health among postdoctoral women *Front. Psych.* **10** 1297

[64] Nagoski E and Nagoski A 2019 *Burnout: The Secret to Unlocking the Stress Cycle* (New York: Ballantine)

[65] Platt J, Prins S, Bates L and Keyes K 2016 Unequal depression for equal work? How the wage gap explains gendered disparities in mood disorders *Soc. Sci. Med.* **149** 1–8

[66] Reilly E, Awad G, Kelly M and Rochlen A 2019 The relationship among stigma, consciousness, perfectionism, and mental health in engaging and retaining STEM women *J. Career Dev.* **46** 443–57

[67] McCullough L 2024 Dear students: class is canceled until further notice while I do my job *McSweeney's* 27 Mar https://mcsweeneys.net/articles/class-is-canceled-until-further-notice-while-i-do-my-job (Accessed: 29 June 2024)

Chapter 9

A history of the topic

We believe that there are few individuals of that gender which plumes itself upon the exclusive possession of exact science, who may not learn much that is both novel and curious in the recent progress of physics from this little volume.

—Rev. William Whewell, in an 1834 review of 'On the connexion of the physical sciences' by Mary Somerville [1]

When did people first start being interested in looking at issues related to women and physics? A long, long time ago. When did we formally start doing research on it? In the last century. In this chapter, I look into different aspects of the history of women's participation in physics and science.

9.1 Educating women in physics

The history of women in science provides some fascinating and startling facts. Women in the US at the turn of the nineteenth century were specifically trained in science, whether it was because 'classics [are] for gentlemen' [2] or because domestic science was a hot topic [3]. In 1910, an MIT professor wrote in *Good Housekeeping* about the usefulness of a conceptual course in physics for home economics curricula [4]. Physics textbooks for women were even used to promote the use of new household technologies [5]! *Physics Today* had a nice overview article on this in 2022 [6].

So why the change in educating women in physics? Blame the Committee of Ten in the US, changing gender expectations after the World Wars, the status of the field, and a host of other reasons. It's all about culture, and culture changes. It's also worth noting that educating high school girls in the expectation they will need this training to become good wives and mothers is not the same as education aimed at welcoming a career woman into the hallowed halls of professional physics. But let's see how we've done over the last century.

9.2 Initiatives of note

The American Physical Society in 1972, after the forming of the Committee on Women in Physics, published a council letter to all employers of physicists, in which they press for equal treatment of women and encouragement of women students [7].

In 1979, the United Nations Conference on Science and Technology for Development passed a single resolution: on the importance of including women in all stages of development projects and programs [8].

The AAAS created a National Network of Minority Women in Science in the 1980s, and at one of their meetings, they noted the extra demands on minority faculty in a majority institution: providing 'the minority's point of view' in meetings and on committees, counseling minority students, and serving as sounding boards for sensitive topics [9]. Today's BIPOC women will empathize.

The US NSF commissioned a short film in 1982 on women and science, including the then-president of AAAS, E Margaret Burbidge, a space physicist [10].

In 1988 the US Congress created the Commission on the Advancement of Women and Minorities in Science, Engineering and Technology Development. The Commission spent nearly two years looking at issues for women, minorities, and people with disabilities and created a report: 'Land of plenty: diversity as America's competitive edge in science, engineering and technology' [11]. Here was a national-level report claiming that diversity was going to save the STEM fields and America's standing in the world.

In 1993 the US National Science Foundation started the Program for Women and Girls (PWG), and in 1998 the name changed to the Program for Gender Equity in STEM. An impact study of the program was conducted in 2000 and found that the 'scope and impact in the field of gender equity and SMET [sic] is unmatched by any other privately or publicly funded program' [12].

The Australian Institute of Physics instituted their Women in Physics Lecturer program in 1997.

The International Union of Pure and Applied Physics (IUPAP) created its Working Group 5: Women in Physics in 1999, and every three years IUPAP sponsors the International Conference on Women in Physics. The IUPAP also has a Gender Champion, and participates in the international Standing Committee for Gender Equality in Science.

9.3 Journals

In the journal *Science*, the archives go back to 1880. Probably the first discussions of women and science are in 1910. In an analysis of the list of American Men in Science of 1910, Cattell includes this sentence: 'There does not appear to be any social prejudice against women engaging in scientific work, and it is difficult to avoid the conclusion that there is an innate sexual disqualification.' They followed up with 'It is possible that the lack of encouragement and sympathy is greater than appears on the surface ...' [13]. Not surprisingly, several readers took umbrage at these statement, and in a subsequent issue one letter makes a wonderful case for differences in the way society treats boys and girls [14], while another argues that if the department chair makes such a difference, the fact that women aren't eligible to be chair should be considered [15]. Both letters note that women are welcome in physics...as lab assistants.

Two weeks after these letters were published, *Science* published a reprint of a *New York Evening Post* piece suggesting that Mme Curie will likely be admitted to the 'charmed circle' of 'the most distinguished group of scientific men in the world'. If she is not, it will be 'one more proof of the immeasurable difference between the degree of encouragement and incentive held out to women and that held out to men' [16].

The first special issue of *Science* on women in science was in 1992, with issues in 1993 and 1994 as well.

In the *American Journal of Physics*, the first occurrence of the word 'woman' is in 1937, with 'Why the woman student does not elect physics' [17]. Daffin surveyed women's colleges with regards to biology, chemistry, and physics enrollments. You can guess that physics was lowest. Shortly after, Sister M Ambrosia offered up 'Teaching physics to women' [18]. She argued that starting with content appealing to women will help, and changing how laboratory work is conducted, along with side notes of blaming men for women's 'inferiority complexes' in physics and suggesting that nothing will help cure this prejudice more than a group of successful, interested women students.

The next significant round of discussing women comes with Robert Romer's 1988 editorial '958 men, 93 women—how many Lise Meitners among those 865?' [19]. Probably inspired by the special issue of *Science* that year [20], he offers a thoughtful commentary. Fascinatingly, he mentions journal editor biases before much research had been conducted on them. Romer also notes that he tries to remove unnecessary gendered language (pronouns), and offers a few suggestions.

This editorial sparked a fascinating round of arguments in subsequent issues. A guest editorial 'Why so few women?' [21] on the current status of the field produced a letter arguing for biological differences, [22] then a three-page response to *that* letter, arguing against the biological conceit and adding more evidence to the social issues [23]. Then another letter from the original guest comment author 'In his response ... Levin imputes to me opinions that I did not express and makes many statements with which I must disagree' [24], a handful more letters to the editor, and two years later, an update [25]. My original file folder on this was labeled 'AJP Letter Wars'.

In *Australian Physicist*, the earliest article is probably September 1975, 'Women in physics: a review of current thinking' [26], but this was a response to part of an editorial in February 1975 [27], which noted that only 3.5% of Australian Institute of Physics (AIP) members were women.

The *Australian Journal of Physics* had its first letter on women in physics in 1980, from Rachel Makinson, then two articles in 1983 [28]. Binney notes that the Makinson letter was the first appearance in the Australian physics community of concern about this issue.

In *Physics Today*, probably the earliest specific article on women in physics is in 1948, by a teacher at a women's college (Bryn Mawr), claiming that female physics students are just as capable as their male peers [29]. After that, it is the eminent Lise

Meitner's address to Bryn Mawr in 1959: 'The status of women in the professions' [30]. It is interesting that the earliest articles in both the *American Journal of Physics* and *Physics Today* are related to women's colleges.

Immediately after Meitner's address, Weeks pulls together a nice bit of data on where women were in physics at that time. She gives the percent of women who were APS Fellows (around 2%), notes APS leaders, and leaders in physics more generally [31]. She also gives a table with degree data: women were less than 5% of degree earners at all levels, and only 1.7% at the PhD level.

Physics World started publication in 1988, and in the very first issue was a note from the Women in Physics Subcommittee about the IOP career break kit [32]. The second volume had an article about women returning to physics after a career break [33].

In *The Physics Teacher*, one of the first articles on gender is an editorial by the same author as the first *Physics Today* article—Walter C Michels [34]. He writes about what has happened for women in physics in the 20 years since his previous essay. Instead of blaming social pressures, he posits that young women find physics courses 'too hard and too discouraging'. I might argue that those *are* social pressures.

A bibliometric analysis of physics education research articles noted that there is significant emphasis on gender equity in the physics education research literature [35]. This is no surprise to those of us in physics education research. Since much of the research literature on women in physics ends up in physics education journals, I also looked at *Physical Review Physics Education Research*. And I found my work done for me! Rachel Scherr did a very nice analysis of papers up to the 'Focused Collection on Gender in Physics' [36]. Of 297 articles published before 2016, 69 mention gender intentionally, and 12 of those actually address a question on gender. Most of the papers on gender reported on performance gaps. Only about 3% of the papers had anything to do with race, ethnicity, disability, or other identities.

Physics Education did a special issue about girls in science, all the way back in 1979. This included a study of exam performance in secondary students; girls from single-sex schools did better than boys, while those in mixed schools did more poorly than boys [37]. Other articles in the issue address textbooks [38], gap data [39], and one is a nice overview of the topic [40].

Physics in Canada was harder to search online. A search for 'women' gave 355 articles, going back to 1957, but most of the earliest ones were not articles specifically about women. The earliest specific article I could identify is from 1984, when W J Megaw wrote about the dearth of women in physics, and its roots in the primary and secondary schools. In a rather damnatory claim, he says he has heard from many students 'who assure me that they have dropped math in grade eight, chemistry in grade seven, and physics probably in kindergarten' [41]. Megaw wrote again in 1992 on gender in physics departments [42]. There was a special theme issue in 1996 on women physicists in Canada, edited by the first female president of CAP/ACP [43].

In 1978, one of the most popular books for science teacher training [44] included this statement in the section about disadvantaged children and language skills:

> Girls at all socioeconomic levels act with respect to science as though they were handicapped. They know less, do less, explore less, and are prone to be more superstitious than boys. [44]

This statement, so appalling today, was thankfully followed up by a bit of a palliative:

> It is tempting to speculate that one reason so little science is being given to the groups who most need it may be related to the feelings of low confidence so many women have when it comes to science. [44]

We have come a long way, right? ... Right?

9.4 Leadership

In the American Association for the Advancement of Science, women had leadership roles early on, starting in 1885. They were in all in anthropology. Several were secretaries and two were chairmen [sic] of the anthropology section [45].

The first woman president of the APS was a very well-known physicist, Chien-Shiung Wu, leading the organization in 1975 [46]. The next was Millie Dresselhaus, president in 1984. Third was Myriam Sarachik in 2003, then Helen Quinn, Cherry Murray, Laura Greene, Frances Hellman, and Young-Kim Lee. Since 1899, eight women as president. But six since 2000—definitely getting better!

The history of the Institute of Physics was difficult to find. A broken link once led to a list of women officers. The IOP was incorporated in 1920, after several decades of a relationship with the Physical Society of London. [47] The Physical Society was open to women from the beginning; I do wonder if any women joined. A note in the 1988 *Physics World* mentions a Women in Physics Subcommittee [32], but I can't find more about that group. There is now a special interest group Women in Physics.

In 1972 the American Physical Society formed the Committee on the Status of Women in Physics. The CSWP is still going strong and manages grants, site visits, and professional development activities.

The Australian AIP as it currently exists was formed in 1963, although it was in 1923 when a Professor A D Ross attempted to start an Australian branch of the IOP, and 1944 when the Australian branch began formal work. AIP created a Women in Physics group in 1994, with Cathy Foley as chair, and in 2007 Cathy Foley became the AIP's first President [28].

Rachel Makinson, a physicist, was the first woman to become Chief Research Scientist at Australia's research organization CSIRO in 1977, and was the first assistant chief of the division in 1979 [48].

The Canadian Association of Physics was formed in 1945, and elected its first woman president in 1993 (A C McMillan), coinciding with an amendment to officially recognize the role of women in physics [49].

9.5 Research

There is very little written about the history of research on women and science. So I will do my best using my experience of two decades of being in this field. In 1979 a fascinating study of 'handicapped' [sic] women scientists was announced in *Science* [50]. One of the earliest studies of this double-minoritized group! A great start, but not really followed up on for a while.

The phrase 'chilly climate' was used by Roberta Hall and Bernice Sandler in a report from the Association of American Colleges' (AAC) Project on the Status and Education of Women, published in 1982 [51]. This report uses the phrase 'micro-inequities', coined in 1977 by Mary P Rowe, Special Assistant to the President at MIT [52] to describe the small behaviors and actions that serve to 'other' someone. She distinguished the term from 'micro-aggressions', which was developed by Pierce in 1974 [53] specifically about racism. The chilly climate report reads in a remarkably familiar manner for the world forty years on. While not focused on science, there is a small section on women in traditionally 'masculine' fields. The authors claim two likely reasons for low participation of women in these areas: departmental climate, and women's own concern over the appropriateness of a 'non-traditional' major; echoes of today's work on climate and identity.

A follow-up report on the chilly climate for faculty in 1986 notes that women faculty in male-dominated fields experience all the problems that women faculty face plus others because of their low numbers, such as increased overt sexism, disparagement of women's abilities, and trouble getting collaborators because of women's perceived lower potential [54]. The AAC (now the American Association of Colleges and Universities) released another report in 1995 by Angela Ginorio: 'Warming the climate for women in academic sciences' [55]. This may be one of the first reports to provide an overview of women in science from high school through degrees to faculty roles. The issues covered are very familiar: the numbers, sexism, confidence, mentoring, harassment, isolation, etc. There is a bit of interdisciplinary work too, noting a few studies on women of color, showing that both racism and sexism are factors.

In my copy of the *Handbook on Research in Science Teaching and Learning* from 1994, the chapter on gender issues goes through achievement gaps, the biological argument, attitudes, stereotyping, family factors, and interventions [56].

A phenomenal resource early on in the literature is Helen Rossiter's two-part work *Women Scientists in America* [57]. She has been lauded for being one of the first people to uncover the work of many women who had disappeared from our history.

9.6 Gender presentation

An interesting cultural piece for women in physics is women's gender presentation. I have not found a specific source for this; however, it's a theme I can easily see from decades of reading in the area. Women in physics have had to choose how feminine they want to look, and that choice has varied over time. Gornick describes one physical chemist as 'a woman who long ago adopted the style and manner of the de-sexed bluestocking' [58]. In wartime Britain, Helen Fraser 'antagonized people by refusing to conform. Instead of the dark, unobtrusive clothes adopted by other women, Fraser favoured elegant, impractical costumes.' [59] Before it was acceptable for women to wear trousers, a woman in physics had to wear dresses. But they could decide what type of dress, as Marie Curie did: 'I have no dress except the one I wear every day. If you are going to be kind enough to give me one, please let it be practical and dark so that I can put it on afterwards to go to the laboratory.' [60] Once women had more flexibility, many women chose more masculine dress, whether it was for practical reasons, to be less noticed, or to partake of some of the masculine culture of physics. Gornick's comment on a biophysicist trained in high-energy: 'Beneath her thick unshaped hair, neat, unadorned features announce: Take no notice of me. Pretend I'm not here' [58]. In the era of third-wave feminism (or are we up to fourth-wave?), women unfortunately still have to deal with the issue of dress as part of their presentation to the world and within their field. See chapter 4 for some of the research on this. A 2006 book with advice for women in science notes that women have to dress more formally than men to achieve the same level of professionalism [61].

9.7 How it's changed

What we see by looking at the history of this topic is a many-decades-long quest to find out why women remain at only 20% of the physics population. My thoughts on the field? Initially, the conversation was all about 'Eeek! There are women in my science!', then we moved on to 'what's wrong with women?' and then 'let's fix the women'. Recently we have finally progressed to 'what's up with physics?', with research about culture and how the culture of physics creates barriers for women. I feel like we are finally in the right place to start making changes that will move our field toward equity.

References

[1] Whewell W 1834 Review of on the connexion of the Physical Sciences by Mrs. Sommerville *Quart. Rev.* **51** 54–68
[2] Tolley K 1996 Science for ladies, classics for gentlemen: a comparative analysis of scientific subjects in the curricula of boys' and girls' secondary school in the United States, 1794–1850 *Hist. Educ. Quart.* **36** 129–53
[3] Dreilinger D 2021 *The Secret History of Home Economics* (New York: Norton)
[4] Norton C L 1910 The relation of physics to home economics courses *Good Housekeeping* **50** 494–6
[5] Behrman J 2017 Domesticating physics: introductory physics textbooks for women in home economics in the United States, 1914–1955 *Hist. Educ.* **46** 193–209
[6] Behrman J 2022 Physics…is for girls? *Phys. Today* **75** 30–6
[7] APS Council 1972 APS Council letter on women *Phys. Today* **25** 62
[8] [no author] 1979 Women, science, and technology at UNCSTD: a follow-up *Science* **206** 1206
[9] [no author] 1980 Minority women in science *Science* **207** 1063
[10] [no author] 1982 Science: woman's work *Science* **217** 44

[11] US NSF 2000 Land of plenty: diversity as America's competitive edge in science, engineering and technology *Report* cawmset0409 Commission on the Advancement of Women and Minorities in Science, Engineering and Technology Development https://www.nsf.gov/publications/pub_summ.jsp?ods_key=cawmset0409

[12] US NSF 2000 Summary report on the impact study of the National Science Foundation's Program for Women and Girls *Report* Urban Institute Education Policy Center https://www.nsf.gov/pubs/2001/nsf0127/nsf0127.pdf p 8

[13] McKeen Cattell J 1910 A further statistical study of American men of science *Science* **32** 672–88

[14] Hayes E 1910 Women and scientific research *Science* **32** 864–6

[15] Talbot M 1910 Eminence of women in science *Science* **32** 866

[16] [no author] 1910 Women and scientific research *Science* **32** 919–20

[17] Daffin J 1937 Why the woman student does not elect physics *Am. J. Phys.* **5** 82–5

[18] Ambrosia S M 1940 Teaching physics to women *Am. J. Phys.* **8** 289–90

[19] Romer R 1988 958 Men, 93 women—how many Lise Meitners among those 865? *Am. J. Phys.* **56** 873–4

[20] Koshland D 1988 Women in science *Science* **239** 1473

[21] Button-Shafer J 1990 Guest comment: why so few women? *Am. J. Phys.* **58** 13–4

[22] Levin M 1990 Women—why so few? *Am. J. Phys.* **58** 905

[23] Ruskai M B 1990 Guest comment: Are there innate cognitive gender differences? Some comments on the evidence in response to a letter from M Levin *Am. J. Phys.* **59** 11–4

[24] Button-Shafer J 1991 Response to 'Women—why so few?' by Michael Levin *Am. J. Phys.* **59** 199

[25] Price J 1993 Guest comment: gender bias in the sciences—some up-to-date information on the subject *Am. J. Phys.* **61** 589–90

[26] Robertson G 1975 Women in physics; a review of current thinking *Aust. Phys.* **12** 122–5

[27] [no author] 1975 Editorial: physics for women *Aust. Phys.* **12** 17

[28] Binnie A 2008 A short history of the AIP Part 3 *Aust. Phys.* **45** 93–9

[29] Michels W 1948 Women… in physics *Phys. Today* **8** 16–9

[30] Meitner L 1960 The status of women in the professions *Phys. Today* **13** 16–21

[31] Weeks D 1960 Women in physics today *Phys. Today* **13** 22–3

[32] Parker B 1988 Contact making and breaking *Phys. World* **1** 47

[33] Jackson D and McCormick E 1989 Women returning to physics *Phys. World* **2** 25–8

[34] Michels W 1968 Editorial: women in physics—two decades later *Phys. Teach.* **6** 283

[35] Süzük E 2023 Physics education research over the last two decades: a bibliometric analysis *Int. J. Educ. Tech. Sci. Res.* **8** 1576–613

[36] Scherr R 2016 Editorial: never mind the gap: gender-related research in *Physical Review Physics Education Research* 2005–2016 *Phys. Rev. Phys. Educ. Res.* **12** 020003

[37] Harding J 1979 Sex differences in examination performance at 16 + *Phys. Educ.* **14** 280–4

[38] Taylor J 1979 Sexist bias in physics textbooks *Phys. Educ.* **14** 277–80

[39] Thompson N 1979 Sex differentials in physics education *Phys. Educ.* **14** 277–80

[40] Ormerod M, Bottomley J, Keys W and Wood C 1979 Girls and physics education *Phys. Educ.* **14** 271–7

[41] Megaw W J 1984 Women in physics *Phys. Canada* **40** 8–9

[42] Megaw W J 1992 Gender distribution in the world's physics departments *Phys. Canada* **49** 25–8

[43] McMillan A C 1996 *Women Physicists in Canada* **52** 94–126

[44] Budd Rowe M 1978 *Teaching Science as Continuous Inquiry* (New York: McGraw-Hill) p 69

[45] Newcombe McGee A 1924 Women officers of the Association for the Advancement of Science *Science* **59** 577

[46] APS presidents by year 1899–1925 *American Physical Society* https://aps.org/about/presidents (Accessed: 19 June 2024)

[47] History of the Institute of Physics https://iop.org/about/iop-history (Accessed: 19 June 2024)

[48] Rachel Makinson *Wikiwand* https://wikiwand.com/en/Rachel_Makinson (Accessed: 19 June 2024)

[49] Ford F 2000 The evolution of CAP/ACP activities *Physics in Canada* **56** 78–86

[50] [no author] 1979 Education and careers of handicapped women scientists studied *Science* **206** 1206

[51] Hall R M and Sandler B R 1982 The classroom climate: a chilly one for women? *Report* Association of American Colleges' Project on the Status and Education of Women https://files.eric.ed.gov/fulltext/ED215628.pdf

[52] Rowe M 1977 The Saturn's rings phenomenon: micro-inequities and unequal opportunity in the American economy *Proc. NSF Conf. on Women's Leadership and Authority* P Bourne and V Parness https://mitmgmtfaculty.mit.edu/mrowe/micro-inequities/ (Accessed: 26 June 2024)

[53] Pierce C 1974 Psychiatric problems of the black minority *American Handbook of Psychiatry* **vol II** ed G Caplan (New York: Basic)

[54] Sandler B and Hall R 1986 The campus climate revisited: chilly for women faculty, administrators, and graduate students *Report* Association of American Colleges' Project on the Status and Education of Women https://files.eric.ed.gov/fulltext/ED282462.pdf

[55] Ginorio A 1995 *Warming the Climate for Women in Academic Sciences* (Washington, DC: American Association of Colleges and Universities Project on the Status and Education of Women)

[56] Gabel D (ed) 1994 *Handbook of Research on Science Teaching and Learning* (New York: Macmillan)

[57] Rossiter M 1992 and 1995 *Women Scientists in America* (Baltimore, MD: Johns Hopkins University Press)

[58] Gornick V 1990 *Women in Science* (New York: Touchstone) p 63; 112

[59] Fara P 2018 *A Lab of One's Own* (Oxford: Oxford University Press) p 207

[60] Marie Curie quotes *BrainyQuote.com* www.brainyquote.com/quotes/authors/m/marie_curie.html (Accessed: 26 June 2024)

[61] Pritchard P (ed) 2006 *Success Strategies for Women in Science: A Portable Mentor* (Burlington, MA: Elsevier) p 174

IOP Publishing

Women and Physics (Second Edition)

Laura McCullough

Chapter 10

The view through rose-colored glasses

A generation later progress for women throughout the academic world is measurably improved .... But other ... forms of inequities linger and obstacles still exist.

—Donna Shalala [1]

Despite the many hurdles and barriers that women face, we do have successful women in physics. Many websites and other venues promote women in physics as role models, and these sites also serve to educate and remind the public that women can do physics.

How have things gotten better for women? What barriers did our mothers and grandmothers face that our sisters and daughters will not? This chapter explores the progress made and the hurdles dismantled for women in physics.

For women raised in the last half of the twentieth century, it was not uncommon to hear that women don't or can't do science. High school girls were advised away from science by teachers and counselors. College professors and advisors were often worse: 'What makes you think you're worth educating? You're a woman, and you're already married.' [2] Today such overt discrimination is rare, though not entirely extinct. Witness the quote in 2015 from Nobel Laureate Tim Hunt [3] about why he doesn't like having 'girls' in the lab. On the plus side, he got a great deal of push back. Even as recently as 2000, it would probably have gone mostly unremarked.

Gender bias in three areas has likely been reduced for women in science: grant funding, journal reviewing, and hiring [4]. The authors of the paper find that given similar resources, women and men do equally well in these areas. These same authors have published other articles [5] claiming either no bias or bias in favor of women, though not without rebuttals [6]. Also, as discussed in earlier chapters, we are not yet at a point where women and men *are* given similar resources.

The idea that biological differences are to blame for women's under-representation has one foot in the grave, and the mountains of literature showing cultural variability and societal bias as the primary movers should be finishing that off soon. I also find the shift from 'gender gaps' to identity and belonging research to be very heartening. There are a growing number of workshops and programs promoting inclusive teaching. We are shifting the culture, one step at a time.

The often-cited article by Wenneras and Wold [7] from 1997 showing gender bias in Swedish grant awards has shed some of its currency in terms of award bias, as Sweden's male–female difference in grant success has been significantly reduced [8]. In physics, we have seen great success coming from a switch to double-blind grant reviewing [9]. We have removed a great deal of the explicit discrimination [10], and things seem to be shifting in a good direction in terms of bias in grants and reviewing.

Any discussion of women's progress in the US must acknowledge the positive effects of Title IX [11] of the Education Amendments of 1972:

> No person in the United States shall, on the basis of sex, be excluded from participation in, be denied the benefits of, or be subjected to discrimination under any education program or activity receiving Federal financial assistance.

Under Title IX, any class, program, or activity that receives federal money cannot deny access because of a person's sex. While the link to physics is less direct than the link to athletics or school programs, Title IX moved the US significantly forward on the path towards gender equality in our schools. This progress has helped promote girls' participation in physics in more ways than can be easily traced back to the legislation. For example, by increasing female participation in sports it may have given them better grounding in some of the contexts used in physics examples that used to favor male contextual knowledge and identification. The Women's Educational Equity Act a few years later had a specific focus on girls' participation in science and math, and funds were set aside in 1991–93 to 'increase the participation of women in instructional courses in mathematics, science, and computer science' [12].

In the UK, the 1988 Education Reform Act [13] served to help girls' education with its focus on maths and science as a core curriculum for all students in every school. As with the American Title IX, though the focus was not specifically on STEM education, it has had a far-reaching impact on how girls were educated in science. Teaching girls science content is necessary. Teaching about successful women in science should also be a part of our curriculum.

One way to learn about and acknowledge the work that women do in physics is to note the groundbreakers, the women who were the first to succeed at certain tasks. The American Institute of Physics has developed a timeline of important events for women in physics and astronomy [14]. We need to acknowledge and honor those who have gone before.

In 1959 Edna Yost wrote: 'Very little biographical material is available about women who are scientists.' [15] Today, there are a great many resources listing successful women in physics and science in general [16]. Biographies abound of famous and less-well-known women who loved physics and made significant contributions. There are lists of books for kids about women scientists [17]. It is now easy to look up and find out interesting facts about female physicists. I believe a day will come when you ask a person to name an important female physicist [18] and they can come up with someone other than Madame Curie.

Another sign of how things have improved is the awareness of women's issues in science in the general public. There are hundreds of books available if one searches for 'women and science'. A web search will find videos, articles, books, blogs, and conferences on the topic. In order to truly solve a problem, one must first recognize that there is a problem. We are solidly at this point now, with more and more general societal awareness and press attention on the issues that women in science face.

An interesting way to look at the general public's view of women in physics comes from the popular US television show *The Big Bang Theory*. Who could imagine that a TV sitcom about physicists would be popular enough to last ten seasons? What great news, right? And yet …. The portrayal of women in science in the show can be quite negative. Shrill. Dowdy. No fashion sense. Admittedly smart, but most often socially awkward. This mirrors the portrayal of male scientists on the show in many ways, but may play differently with the potential audience of young women interested in science. Especially with a pretty, socially adept woman who is not a scientist as a figure of comparison in the show. Are these role models that our young women will connect with, or aspire to become?

Issues of women and physics have been recognized by our professional organizations for many years. Most of the professional organizations for physicists in the English-speaking world have a committee or a special topical group on women and/or diversity in physics (one of these committees has been around since 1972!) [19]:

- Diversity and Equity Group in Australian Physics (Australian Institute of Physics) https://www.aip.org.au/Diversity-and-Equity-Group-in-Australian-Physics-(DEGAP)
- Inclusive physics (American Physical Society) https://www.aps.org/initiatives/inclusion
- Committee on Women in Physics (American Association of Physics Teachers) https://www.aapt.org/aboutaapt/organization/women.cfm
- Diversity and inclusion (UK Institute of Physics) https://www.iop.org/about/IOP-diversity-inclusion
- Equality, diversity and inclusion (EDI) (UK Research and Innovation) https://www.ukri.org/what-we-do/supporting-healthy-research-and-innovation-culture/equality-diversity-and-inclusion/
- WG5: Women in physics (International Union of Pure and Applied Physics) https://iupap.org/who-we-are/internal-organization/working-groups/wg5-women-in-physics/

Many of the physics professional organizations not only have statements about diversity and inclusion, but they are also making policies to support these goals. Codes of conduct are becoming the norm. Sexual harassers have been disinvited from speaking gigs and banned from some institutions [20]. The #MeToo movement has increased public awareness of sexual harassment and abuse. Hashtags such as #ThisIsWhatAScientistLooksLike and the If/ThenSheCan project are making it easy to see women as scientists.

There is also a growing amount of encouragement and support for allies, specifically male allies [21]. At the Division of Nuclear Physics conferences, there are trained allies who wander the conference looking for people who are uncomfortable or experiencing harassment [22]. They are the go-to for people who want to report an incident.

Bystander intervention is also becoming a popular topic for workshops. While we don't have a lot of research yet on its overall effectiveness in reducing harassment, it does help the targeted individual [23]. The ADVANCEGeo program has a great workshop for scientists [24].

I find a lot of hope in our younger people. The Gen Z [25] generation is one of the most inclusive we have had, with more racial diversity and more comfort with non-binary gender categories, as well as more acceptance of LGBT+ identities [26]. In an evaluation I did of grants for women in physics groups, quite a few groups were changing their names to Women+, Underrepresented People, Gender Minorities, etc. The recent changes from 'Conferences for Undergraduate Women in Physics' to 'Conferences for Undergraduate Women and Gender Minorities in Physics' (in the US) and 'Conference for Undergraduate Women and Non-binary Physicists' (CUWIP+ in the UK) are another change that keeps me hopeful.

I also find hope in the number of women who especially enjoy challenging the stereotype of physicist = male. 'I kind of enjoy the taboo that was around it, not many people did physics, and then even more so not many girls did physics. I think, I kind of enjoyed that.' [27] I have heard this from colleagues as well as reading about it in the literature. You go, girl!

Initiatives such as Project JUNO, LimitLess, SEA CHANGE, and Improving Gender Balance show that national organizations take this issue seriously and are putting resources behind it. The Australian Institute of Physics supports a Women in Physics Lecturer program [28]. APS funds multiple programs, including the APS-IDEA network, climate site visits, and grants for women in physics groups. The APS and the AAPT supported a program called Departmental Action Leadership Institutes, helping departments build more inclusive cultures. While not a panacea, the research on DALI is a great starting point on exploring how best to get departments to shift their culture [29]. Culture does change, though slowly, and often against great resistance.

Most of these programs and initiatives did not exist thirty years ago, and their presence today suggests that we have indeed come a long way. Though we still have under-representation of women in physics, it is worth celebrating our progress as we reach for more. Things continue to get better. Slowly, but surely.

We are now twenty years past the time when then-Harvard-president Larry Summers claimed that the three likely reasons women aren't in science are (i) status related ('high-powered job'), (ii) differential ability at the high-aptitude end, and (iii) patterns of discrimination [30]. His remarks set off a national and international furor. The number of women in physics careers has improved very little since then, though the research brings much hope. Unfortunately, it's still a hot-button issue with a lot of sexism surrounding it. In 2014 Neil deGrasse Tyson was asked at a conference: 'The Larry Summers question: What's up with chicks and science?' [31]. Tyson responded well, speaking about bias and assumptions. But just the phrasing of the question is enough to make many people flinch. That anyone could use such language in our modern world in a public forum is a perfect example of the continuing need to fight sexism.

Despite the numerous problems and barriers that women in physics face, it is reassuring to know that we have made many gains in the last century. While the 'problem' of women in science is not yet solved, there are many things that today's women don't need to deal with that our predecessors did.

So that's the case for how things are better; what are the barriers and hurdles that remain? What do we still need to work on? The next chapter looks at some of the other issues currently being faced by female physicists, and women in general.

References

[1] Shalala D 2009 *Women in Science: Then and Now* ed V Gornick (New York: Touchstone) p vii
[2] Gornick V 2009 *Women in Science: Then and Now* (New York: Touchstone) p 94
[3] Chappell B 2015 Nobel Laureate in hot water for 'trouble with girls' in labs *The Two-Way* 10 Jun www.npr.org/sections/thetwo-way/2015/06/10/413429407/nobel-laureate-in-hot-water-for-trouble-with-girls-in-labs (Accessed: 13 June 2024)
[4] Ceci S and Williams W 2007 *Why Aren't More Women in Science?* (Washington, DC: American Psychological Association)
[5] Ceci S and Williams W 2015 Women have substantial advantage in STEM faculty hiring, except when competing against more-accomplished men *Front. Psychol.* **6** 1532
Ceci S, Ginther D, Kahn S and Williams W 2015 Women in science: the path to progress *Sci. Am. Mind* **26** 62–9
[6] Guy S 2015 Too good to be true? *SWE Mag.* **61** 18–20
Bernstein R 2015 Women best men in study of tenure-track hiring *Science* **348** 269
[7] Wenneras C and Wold A 1997 Nepotism and sexism in peer-review *Nature* **387** 341–3
[8] Sandstrom U and Hallsten M 2008 Persistent nepotism in peer-review *Scientometrics* **74** 175–89
[9] Strolger L and Natarajan P 2019 Doling out Hubble time with dual-anonymous evaluation *Phys. Today* 1 Mar
Singh Chawla D 2021 Record number of first-time observers get Hubble telescope time *Nature News* 25 Nov
[10] Ceci S, Ginther D, Kahn S and Williams W 2015 Women in science: the path to progress *Sci. Am. Mind* **26** 62–9
[11] 20 US Code § 1681—Sex https://uscode.house.gov/view.xhtml?req=20+USC+1681 (Accessed: 13 June 2024)
[12] White K D, Mulhauser F V, Bogart G H, Chennareddy V and Salzman J 1994 Women's Educational Equity Act: a review of program goals and strategies needed *Letter Report* GAO/PEMD-95-6 Government Accountability Office https://govinfo.gov/content/pkg/

GAOREPORTS-PEMD-95-6/html/GAOREPORTS-PEMD-95-6.htm (Accessed: 13 June 2024)

[13] Education Reform Act 1988 www.legislation.gov.uk/ukpga/1988/40/contents (Accessed: 13 June 2024)

[14] Timeline of women's achievements in physics, astronomy and related disciplines since the 18th century *American Institute of Physics, Center for History of Physics* www.aip.org/sites/default/files/history/files/Timeline-WomenInPhysicsAstronomy.pdf (Accessed: 13 June 2024)

[15] Yost E 1959 *Women of Modern Science* (New York: Dodd, Mead and Co) p vii

[16] Byers N and Williams G 2010 *Out of the Shadows: Contributions of Twentieth-Century Women to Physics* (Cambridge: Cambridge University Press)
Hartline B and Li D 2002 *Women in Physics: The IUPAP International Conference on Women in Physics* (Melville, NY: American Institute of Physics)
Gibson K 2014 *Women in Space* (Chicago, IL: Chicago Review Press)
Rossiter M 1982 *Women Scientists in America* (Baltimore, MD: Johns Hopkins Press)

[17] 21+ children's books about women scientists *Science Books for Kids* http://books.growing-withscience.com/2014/03/24/21-childrens-books-about-women-scientists/ (Accessed: 13 June 2024)
29 children's books about female scientists *Feminist Books for Kids* https://feministbooks-forkids.com/childrens-books-about-female-scientists/ (Accessed: 13 June 2024)

[18] Contributions of 20th century women to physics http://cwp.library.ucla.edu/ (Accessed: 13 June 2024)

[19] Committee on the Status of Women in Physics *American Physical Society* www.aps.org/about/governance/committees/cswp/index.cfm (Accessed: 13 June 2024)

[20] Koren M 2018 Lawrence Krauss and the legacy of harassment in science *The Atlantic* 24 Oct https://theatlantic.com/science/archive/2018/10/lawrence-krauss-sexual-misconduct-me-too-ari-zona-state/573844/ (Accessed: 13 June 2024)

[21] Feldman H 2023 Being an ally *AIP Conf. Proc.* **3040** 060011

[22] DNP Allies Program *American Physical Society* https://engage.aps.org/dnp/governance/about/allies-program (Accessed: 25 June 2024)

[23] Jennings L, Zhao K, Faulkner N and Smith L 2024 Mapping bystander intervention to workplace inclusion: a scoping review *Human Res. Manag. Rev.* **34** 1–21

[24] ADVANCEGeo Workshops https://serc.carleton.edu/advancegeo/workshops/index.html (Accessed: 25 June 2024)

[25] Generation Z *Wikipedia* https://en.wikipedia.org/wiki/Generation_Z (Accessed: 13 June 2024)

[26] Parker K and Igielnik R 2020 On the cusp of adulthood and facing an uncertain future: what we know about Gen Z so far *Report* Pew Research Centre https://pewresearch.org/social-trends/2020/05/14/on-the-cusp-of-adulthood-and-facing-an-uncertain-future-what-we-know-about-gen-z-so-far-2/ (Accessed: 13 June 2024)

[27] Eren E 2021 Exploring science identity development of women in physics and physical science in higher education *Sci. Educ.* **30** 1131–58 p 1142

[28] Women in Physics Lecture Tour 2024 *Australian Institute of Physics* https://aip.org.au/Women-In-Physics-Lecture-Tour (Accessed: 13 June 2024)

[29] Sachmpazidi D, Turpen C, Petrella J, Dalka R and Abdurrahman F 2024 Recognizing dominant cultures around assessment and educational change in physics programs *Phys. Rev. Phys. Educ. Res.* **20** 010132

[30] [no author] 2005 Harvard President Summers' remarks about women in science, engineering *PBS News* 22 Feb https://pbs.org/newshour/science/science-jan-june05-summersremarks_2–22 (Accessed: 29 June 2024)

[31] Neil Degrasse Tyson on being Black, and women in science *D- Angel* (video) www.youtube.com/watch?v=z7ihNLEDiuM (Accessed: 29 June 2024)

IOP Publishing

Women and Physics (Second Edition)

Laura McCullough

Chapter 11

The glass is half-empty

I have always believed that contemporary gender discrimination within universities is part reality and part perception, but I now understand that reality is by far the greater part of the balance.

—Charles Vest, when president of MIT [1]

We have many examples of productive and successful women in physics; things have definitely gotten better. And yet, the percentage of women in physics in Anglophone countries currently continues to hover around 20%. Why is this number not increasing? What is keeping more girls and women from choosing physics, and staying in physics? Some of the ongoing challenges and barriers to women in physics will be explored in this chapter.

A good overview of the issues women face comes from an international survey of physicists [2], where 3000 women from around the world shared what helps and hurts them as physicists. Access to resources, demands from family and home, chances to travel to conferences; all these affect how successful women are in physics. Many of these have been addressed in previous chapters. Here are a few other issues.

11.1 The nature of science

There is an active field exploring the idea of a feminist or masculinist nature of science. Is science as we have created it an inherently male field? Examples abound in the literature of ways that science shows off its masculine side. The first atomic bombs were named Fat Man and Little Boy [3], an example of the idea of male birth analogies in science, argues Carol Cohn [4]. Many histories of science and scientists speak of science as 'a woman to be unveiled, unclothed, and penetrated by masculine science' [5]. Nature is a woman, a harsh mistress, something one falls in love with, even marries. To quote Richard Feynman: 'That was the beginning, and the idea seemed so obvious to me and so elegant that I fell deeply in love with it. And, like falling in love with a woman, it is only possible if you do not know much about her, so you cannot see her faults. The faults will become apparent later, but after the love is strong enough to hold you to her.' [6] Helen Longino has argued that science cannot be truly objective since it is always undertaken in a social setting, and so we should make sure that the biases and values we attach to science are not purely masculine in nature [7].

11.2 Climate for gender and diversity work

Writing this edition of the book in 2024, I must note changes to the climate for women and diversity generally in the US and other countries. Over half of US states have recently introduced or passed legislation restricting any sort of equity, diversity, or inclusion work for state institutions [8]. Women and Gender Studies departments are being de-funded and closed, diversity offices are being shut down, and programs to support under-represented groups are unable to do any work [9]. Policies like these can cause faculty to leave states enacting them [10]. This appalling movement must be stopped and reversed if nations are to maintain any sort of STEM workforce.

There has also been a surge of anger towards programs and events that focus on women and empower women. Much of the anger surrounds the use of the word 'privilege'. Many men feel defensive when told they have privilege, as do White people, and straight people, and able-bodied people. I can understand it, a little. If you have never thought about it, you have likely never thought about ways your life is made easier because you are a White male. You haven't seen it; you haven't noticed your life being particularly 'easy'.

One of the best ways I've seen of talking about privilege comes from John Scalzi, a well-known author and member of the science fiction community [11]. He uses a videogaming analogy: 'In the role playing game known as The Real World, 'Straight White Male' is the lowest difficulty setting there is.' It's a great essay.

11.3 The old 'women are worse at math/science/ spatial reasoning because of biology' myth

A discussion of women and science cannot be complete without discussion of the pernicious idea of inherent biological weakness of women in the area of science. Most of the time this idea is brought up with the research on gender differences in spatial reasoning (see Benbow and Stanley [12] for a classic paper on this topic). Modern thinking on this idea [13] is that there are brain-based differences in how men and women think and solve problems, but that these differences do not limit how well men or women can do science. Nor do we know if these differences are biological and/or sociological in origin. A short training intervention can raise the spatial abilities of both boys and girls significantly [14], suggesting that biology isn't the only factor at play. Our brains are remarkably plastic. Data showing that jugglers [15] and London cabbies [16] have actual changes in size of certain brain areas also suggest that brain/spatial interactions are a two-way street.

Another argument against the idea of women being biologically disadvantaged when it comes to science is the fact that in some societies women make up a much more significant proportion of physicists. Majority Muslim countries often have a higher proportion of women in undergraduate physics than men; a cultural data point that is fascinating [17]! If a double-X chromosome were to blame for the lack of women in science, we would expect very small differences across the globe in women's participation in physics. Instead, we see striking differences between cultures [18].

As the low number of women in physics in the English-speaking world suggests, we still have issues to fight and cultural biases to erode. And yet we have come a long way from the first days where women were not even recognized for their work in the lab, if they were allowed in at all.

References

[1] Vest C 1999 Massachusetts Institute of Technology: a study on the status of women faculty in scienceat MIT *MIT Faculty Newsletter* **11** March http://web.mit.edu/fnl/women/women.html (Accessed: 29 June 2024)

[2] Ivie R, Czujko R and Stowe K 2002 Women physicists apeak *AIP Report* www.aip.org/statistics/reports/women-physicists-speak (Accessed: 29 June 2024)
Ivie R and Guo S 2006 Women physicists speak again *AIP Report* www.aip.org/statistics/reports/women-physicists-speak-again (Accessed: 29 June 2024)
Ivie R and Tesfaye C L 2012 Women in physics: a tale of limits *Phys. Today* **65** 47–50

[3] Little Boy and Fat Man *Atomic Heritage Foundation* www.atomicheritage.org/history/little-boy-and-fat-man (Accessed: 29 June 2024)

[4] Cohn C 2009 Sex and death in the rational world of defense intellectuals *Women, Science and Technology* ed M Wyer *et al* (New York: Routledge)

[5] Fee E 1986 Critiques of modern science *Feminist Approaches to Science* ed R Bleier (Elmsford, NY: Pergamon) p 44

[6] Feynman R 1965 The development of the space-time view of quantum electrodynamics *Nobel Lecture* 11 Dec www.nobelprize.org/nobel_prizes/physics/laureates/1965/feynman-lecture.html (Accessed: 29 June 2024)

[7] Longino H 1990 *Science as Social Knowledge* (Princeton, NJ: Princeton University Press)
Keller E F 1985 *Reflections on Gender and Science* (New Haven, CT: Yale University Press)
Wyer M, Barbercheck M, Giesman Cookmeyer D, Ozturk H and Wayne M (ed) 2009 *Women, Science and Technology* (New York: Routledge)

[8] Kratz J 2024 Anti-DEI legislation: how we got here with tips to combat it *Forbes* 9 June https://forbes.com/sites/juliekratz/2024/06/09/anti-dei-legislation-how-we-got-here-with-tips-to-combat-it/ (Accessed: 14 June 2024)

[9] Asmelash L 2023 DEI programs in universities are being cut across the country. What does this mean for higher education? *CNN* 14 Jun https://cnn.com/2023/06/14/us/colleges-diversity-equity-inclusion-higher-education-cec/index.html (Accessed: 14 June 2024)

[10] Quinn R 2024 Are professors really fleeing universities in red states? *Inside Higher Ed* 3 Jan https://insidehighered.com/news/faculty-issues/tenure/2024/01/03/are-professors-really-flee-ing-universities-red-states (Accessed: 14 June 2024)

Contreras J 2023 'I'm not wanted': Florida universities hit by brain drain as academics flee *The Guardian* 30 Jul https://theguardian.com/us-news/2023/jul/30/florida-universities-col-leges-faculty-leaving-desantis (Accessed: 14 June 2024).

Rahman K 2023 Florida combats colossal teacher shortage *Newsweek* 12 Apr https://newsweek.com/florida-combats-colossal-teacher-shortage-1793928#:~:text=Ron%20DeSantis%20Said%20About%20Education%3F&text=Florida%20has%205%2C294%20teacher%20vacan-cies (Accessed: 14 June 2024)

[11] Scalzi J 2012 Straight white male: the lowest difficulty setting there is *Whatever* 15 May https://whatever.scalzi.com/2012/05/15/straight-white-male-the-lowest-difficulty-setting-there-is/ (Accessed: 14 June 2024)

[12] Benbow C P and Stanley J C 1980 Sex differences in mathematical ability: fact or artifact? *Science* **210** 1262–4

[13] Ceci S and Williams W 2007 *Why Aren't More Women in Science?* (Washington, DC: American Psychological Association)

Xie Y and Shauman K 2003 *Women in Science: Career Processes and Outcomes* (Boston, MA: Harvard University Press)

[14] Applebee D, Bennett-Day B, Ferrari J, Pritchard P and Boettger-Tong H 2021 Multimodal training improves spatial reasoning skills in female college students *J. Sci. Educ. Tech.* **30** 539–49

Stieff M, Dixon B L, Ryu M, Kumi B C and Hegarty M 2014 Strategy training eliminates sex differences in spatial problem solving in a STEM domain *J. Educ. Psychol.* **106** 390–402

[15] Malik J, Stemplewski R and Maciaszek J 2022 The effect of juggling as dual-task activity on human neuroplasticity: a systematic review *Int. J. Environ. Res. Public Health* **19** 7102

[16] Maguire W K 2011 Acquiring 'the Knowledge' of London's layout drives structural brain changes *Curr. Biol.* **21** 2109–14

[17] Moshfeghyeganeh S and Hazari Z 2021 Effect of culture on women physicists' career choice: a comparison of Muslim majority countries and the West *Phys. Rev. Phys. Educ. Res.* **17** 010114

[18] Barinaga M 1994 Surprises across the cultural divide *Science* **263** 1468–72

Ivie R, Czujko R and Stowe K 2002 Women physicists speak: the 2001 International Study of Women in Physics *AIP Report* www.aip.org/statistics/reports/women-physicists-speak (Accessed: 29 June 2024)

Chapter 12

Closing thoughts

> Until a change of the reference system to identify 'who is a physicist' occurs, physics is bound to lose women+ talents.
> —Tracey Berry and Saskia Mordijck [1]

Women's interest in science stretches back to Hypatia and beyond. Women have always cared about the way the world works, whether that interest was expressed in a pursuit called natural philosophy or through the more modern STEM terminology. What they have not always been allowed to do is to follow that passion. While in the world today we are seeing more and more opportunities for girls and women to learn and do science, the road is long, and the end is not yet clear in our sight. Where chemistry, math and biology are nearing gender parity, physics and engineering remain stubbornly male-dominated [2].

When I walked into my physics graduate school on day one and there were twenty-four men and me, I knew that we had a problem. A problem begging for a solution, and because I am a scientist and what I do is solve problems, that moment was the beginning of what has been 25 years of research on gender issues in science. I don't know all the answers, and I doubt the problem will be solved in my lifetime, but I know more than I knew then, and sharing that is part of the solution. Hence this book.

From the early days of believing that women's education would cause their uteruses to shrivel up and disappear, to letting women into some labs but not giving them credit, to saying that women aren't in science because they aren't biologically capable of it, to today's issues of a sense of belonging and identity—we have already traveled a long road. Many of the 'hot' women and science topics of the 1980s are antiquated relics to today's researchers. We can only hope that the issues women face in physics today are just as unimaginable and infuriating to our granddaughters as the idea of telling a little girl that 'physics is only for boys' is to us.

The women and physics researchers of today and tomorrow need to consider how we go about doing our research as well as how a shifting world will require us to keep shifting our focus to move with the times. We should not view men as *the* standard against which women ought to be judged. Instead, we must determine what is best for physics, and move our field in that direction. As Whites are heading towards becoming a minority in many English-speaking countries, the interplay of gender, race, and other identities must become a greater part of our conversation. We need to address the nature of gender as nonbinary. There is also a great deal to be learned by considering the many identities we all hold. The experience of a White straight cis woman is not the same as that of a trans woman, a Latina, or a lesbian in physics. We need to study the experience of women in a way that addresses *all* of their identities.

Culture is hard to change and slow moving when it does. And yet the evidence is clear that cultural factors are the main culprits in the lower participation rates of women in science, which means the culture will have to be moved and we will be the ones who have to move it. As researchers continue to study this complicated and fascinating issue, we learn more and more that we couldn't have imagined even twenty years ago. For example, who would have guessed that the belief that a field requires innate ability or brilliance is strongly correlated with the participation of women in that field? The mere fact that physicists believe that physics requires an innate talent has a strong relationship with physics' low numbers of women [3]. What happens if we can change that belief in inherent talent? We don't know, but we should strive to find out.

We now have a history in this area that includes decades of research on women and physics, the growth of programs supporting girls and women in science, and a real societal awareness that this is an important issue. All of this points to a rosier future, both for women and for the field. I am optimistic that we can and will solve our remaining problems, because I am a scientist and what we do is solve problems, *and* because I am a woman in science, we have always been here, and we always will be.

References

[1] Berry T and Mordijck S 2024 Wasted talent: the status quo of women in physics in the US and UK *Commun. Phys.* **7** 1–2
[2] NSF 2023 Diversity and STEM: women, minorities, and persons with disabilities *Report* National Center for Science and Engineering Statistics (NCSES), Directorate for Social, Behavioral and Economic Sciences https://ncses.nsf.gov/pubs/nsf23315/ (Accessed: 17 June 2024)
[3] Leslie S-J, Cimpian A, Meyer M and Freeland E 2015 Expectations of brilliance underlie gender distributions across academic disciplines *Science* **347** 262–5

www.ingramcontent.com/pod-product-compliance
Lightning Source LLC
Chambersburg PA
CBHW080243270326
41926CB00020B/4351